nowledge.　知識工場
Knowledge is everything！

知識工場
Knowledge is everything！

知識工場
nowledge.
Knowledge is everything！

知識工場
Knowledge is everything！

10秒
一貼
不用抄！

超人氣

Just One **Click!**
Writing A **Perfect E-mail**.

商用英文
E@mail
立可貼大全

張翔
——著

最強「無腦級」偷吃步，在商場打滾的你一定要會，
只要動動滑鼠，複製＋貼上，就能成為職場紅人！

1 確認E-mail主題 *Checked!*

3 替換英文句
Change!

2 貼上範本 *Paste!*

專業職人都懂這一味！
聰明省力的最強無腦剪貼術。

① 速成！How fast

主文範本 10 秒完成

選定要寫的主題之後，即能搭配本書附贈的光碟，Ctrl + C（複製）& Ctrl + V（貼上），10 秒寫 E-mail，效率秒殺眾人。

② 秒懂！Key words

E-mail 關鍵字一目了然

每篇 E-mail 後面的「E-mail 關鍵字一眼就通」，列出這封 E-mail 的重點單字，對照方便，還能擴充英語單字力。

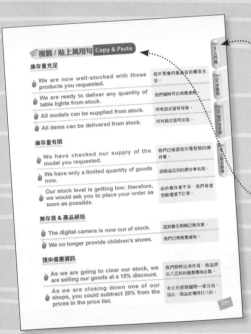

③ 統包！All in one

四大主題 K.O 商用 E-mail

本書包含「入門篇」、「求職篇」、「客戶往來篇」，以及「人際互動篇」4 大章。需要職場英文的你，靠這一本就行。

④ 百搭！Way to go

什麼都能表達的替換神句

每篇 E-mail 之下補充「複製／貼上萬用句」。貼上範本之後，再替換句子，什麼內容都能寫。

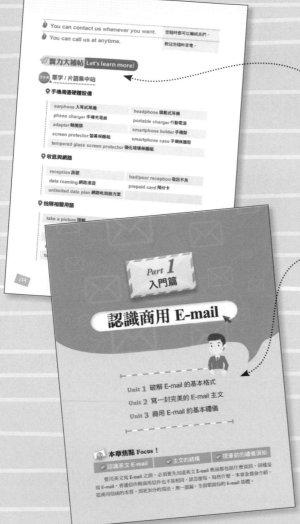

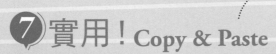

5 全面！Level up

單字／文法／資訊三冠王

每篇 E-mail 最後的「實力大補帖」之下，分成「單字／片語集中站」、「文法／句型解構」，以及「資訊深度追蹤」，英語力全方位並進。

6 扎實！Basics

全面破解商務 E-mail

本書除了實用度破表的 E-mail 範本之外，Part 1 還特別介紹商用 E-mail 的結構，大幅度強化基礎觀念。

7 實用！Copy & Paste

超值資料片大放送

隨書贈送的超值資料片內含所有「E-mail 範本」&「複製／貼上萬用句」。資料片裡的 Word 檔，其檔名規則為「本書的頁數 & 標題」，方便讀者對照使用。

一鍵複製＋貼上，
再也不用絞盡腦汁寫 E-mail ！

學英文，不外乎「聽、說、讀、寫」四大領域，雖然身為教師，常常提醒學生這四大領域要並進學習，不能偏廢，但對學生而言，要將英文化做語言或文字的「說和寫」，往往是最令人緊張的部分。

在從事教學的過程中，我經常會鼓勵學生不要把英文想得太困難，尤其不要在學習之前就一直暗示自己英文很困難（說起來是很理所當然的道理，但請相信我，有不少學生都會不自覺地這麼做）。每每看到學生從害怕英文開始，到能肯定自己，愈學愈快樂的時候，我就能得到莫大的滿足，這或許是我一直無法離開教育這個領域的原因吧。

不過，儘管有許多學生在英文科目上取得成功，但進入職場之後，有一塊領域他們會特別陌生，就是「商務 E-mail」的寫法。英文作文的訓練，我們從國中開始就有，自傳的寫法，申請大學時也會遇到，但是，真的要用英文與客戶溝通的時候，許多人反而不知道該從何下筆。這一點其實非常好理解，我們從小開始，所學的英文都是最基本的構成，加上多數人學英文為的是能與老外溝通，所以並不重視「商務英文」的說法，這一點其實非常吃虧。

雖然同樣是英文，但商務 E-mail 卻是非常特別的一塊領域，除了普通我們知道的「打招呼」用語之外，與客戶往來時，其實更需要拿捏文字。要如何用文字表達自己公司的立場和底線，卻又不過於冷硬？當商品出問題時，要如何堅守立場，卻又不失禮？甚至於，客戶追究起責任時，你要如何藉由一封 E-mail，讓對方感受到誠意，進而願意平心靜氣地對談？這些商務往來常見的情況，偏偏並非日常用語中常接觸的英文。因此，一寫起商務 E-mail，大家往往糾結於單字 & 口氣，這封信會不會太有禮貌導致自己看起來矮一階？這樣寫會不會太過生硬而讓對方以為沒有轉圜餘地？糾結於這些點的同時，最懊惱的 「自己無法判斷筆下的英文究竟會不會招人誤解」。

為了解決這個問題，我才決定編寫這本書。我編寫的目的很簡單，就是為了讓大家「以最輕鬆的方式寫完 E-mail」。在與編輯討論時，我甚至提出「不用腦也能寫 E-mail」的想法，想最大限度地幫助讀者。最後產生「複製＋貼上完成 E-mail」的結構。

不過，也因為這個原因，在編寫目錄時就必須花很多心思。內容是要幫助大家寫商務 E-mail 沒錯，但如果對英文 E-mail 沒有基本的認識，那麼大家使用這本書或資料光碟時，或許會感到疑惑。因此，在重要的 E-mail 內文之外，我編寫了 Part 1 的基本介紹，從 E-mail 的架構，到最基本的禮儀（商務 E-mail 並非朋友間的聯繫，所以尤其必須注意），如果大家花時間閱讀的話，相信會受益良多。

這本書最大的特點，是「複製／貼上完成一封 E-mail」，不過，我希望大家在使用之前，能花點時間看一下 Part 1 的入門篇。在這章的內容中，我除了介紹 E-mail 的結構之外，也花了篇幅去提點大家，下筆寫商務 E-mail 前該注意的事情。網際網路的便利，讓大家習慣於短訊息的發送，但還是得提醒大家，商務 E-mail 並非日常談天，不能以過於輕鬆的口吻寫（否則會讓對方以為你態度隨便，進而影響商務合作），所以，基本的禮儀我也寫進了 Part 1，希望能在幫助大家寫的同時，也強化大家的基本概念。像是不少人喜歡用驚嘆號來表示情緒，但是在商務往來中，過多的驚嘆號只會讓對方覺得你情緒不穩（甚至以為你口氣差），所以在不確定這麼用是否合適之前，建議大家以最平實的逗點和句點完成內容即可。

除了上述提到的內容之外，Part 2 一開始，我就編寫了好幾篇的履歷表＆自傳。雖然實際上的內容要依照讀者本身的情況去修改，但我盡量於書中提示大家要點，相信大家看完這本書，就能以較輕鬆的心態去面對商務 E-mail。

張翔

目錄

Part 3 【客戶往來篇】
與廠商的專業互動

目錄

Contents

目錄

@ Part 4 【人際互動篇】
拉近彼此的關係

Unit 1 ▸ 辦公室點滴

Unit 2 ▶ 祝賀函

Unit 3 ▶ 邀請函

目錄

Part 1 入門篇

認識商用 E-mail

Unit **1** 破解 E-mail 的基本格式

Unit **2** 寫一封完美的 E-mail 主文

Unit **3** 商用 E-mail 的基本禮儀

 本章焦點 Focus！

✓ 認識英文 E-mail	✓ 主文的結構	✓ 提筆前的禮儀須知

　　要用英文寫 E-mail 之前，必須要先知道英文 E-mail 裡面都包括什麼資訊。同樣是寫 E-mail，普通信件與商用信件也不甚相同，該怎麼寫、寫些什麼，本章全都會介紹，從商用信函的本質，到更加分的寫法，無一遺漏，全面鞏固你的 E-mail 基礎。

Unit 1

破解 E-mail 的基本格式
Getting to Know E-mails

From:	Tina Yang <tinayang@ntu.edu.tw>
To:	Lauren Lee <lauren.L123@ntu.edu.tw> **1**
Cc:	Simon Chou <simonc@onextbook.com> **2**
Bcc:	Susan Lin <susanlin@onextbook.com> **3**
Attachment:	2017 Annual Report.pdf **4**
Subject:	2017 Annual Report **5**

Dear Mr. Lee, **6**

Attached is the annual report which I mentioned in our previous call. Kindly give me your suggestions and feedback after reading it.

As your feedback and advice are of great importance and eagerly needed, I would be truly appreciated it if I am able to collect them by the end of this week.

If you have any further inquiries, please do not hesitate to let me know. **7**

Best regards, **8**

Tina Yang **9**

Professor of Department of Foreign Languages and Literatures **10**
National Taiwan University **11**
No.1, Sec. 4, Roosevelt Rd., Da-an Dist., Taipei, Taiwan, R.O.C. **12**
Tel：(02) 2123-4567 **13**
Fax：(02) 2123-4568 **14**
E-mail：tinayang@ntu.edu.tw **15**
Website: www.front-page.com **16**

E-mail 欄位介紹

1 Recipient / Receiver 收件者

2 Carbon copy 副本

3 Blind carbon copy 密件副本

4 Attachment / Enclosure 附件檔案

5 Subject 主旨

6 Salutation 稱謂語

7 Body of E-mail 信件主文

8 Complimentary close 結尾敬語

9 Signature 署名

10 Job position/title 職稱

11 Company name 公司名稱

12 Company address 公司地址

13 Company telephone number 公司電話

14 Company fax number 公司傳真

15 Contact information 聯絡資料

16 Company website address 公司網站

1. Recipient / Receiver 收件者

電子郵件的收件者欄位不像紙本商業書信需要註明頭銜，只需輸入對方的電子郵件位址即可。如果您有定時儲存聯絡人的電子郵件資料，部分電子郵件信箱軟體會在您輸入收件者姓名時，自動顯示對方的電子郵件，此時只需選取即可。

2. Carbon copy 副本

Cc 為 Carbon copy 之縮寫。一份 E-mail 信件，若想同時發送給其他人，讓收件者以外的人了解相關事項，則可於此欄位中輸入其他人的電子郵件，原收件者以及副本收件者皆能得知此封信被發送給哪些人，回信時也可同時寄回給副本收件者。

3. Blind carbon copy 密件副本

Blind 的字義為「看不見」，用在信件上，則為「地址未寫清楚」之意。如果不想讓收件者知道該信件另有副本收件者，則可將副本收件者的電子郵件填於本欄。需要注意的是，若您被列為密件副本的收件者，建議不要直接回覆該信件，以免讓主要收件者得知另有其他收件人的存在。

4. Attachment / Enclosure 附件檔案

為給予收件者更詳細的信件相關資訊，可隨信附上檔案。但送出前務必再次檢查附件，以免寄出錯誤的檔案。此外，為避免對方電腦中毒，寄出前最好先用掃毒軟體檢查檔案；如果收到的信件中有附加檔案，也最好在下載前進行病毒掃描，以降低電腦中毒的風險。

5. Subject 主旨

信件主旨是為了幫助收件者對於信件內容一目了然而下的標題，功能如同作文題目與報紙頭條。一般在職場上，大家必須處理或溝通的信件都非常多，若信件主旨能提綱挈領、簡單扼要，就能節省收件者的處理時間，提高工作效率。

📍 商務類主旨書寫範例

01 Price List 價目表

02 Making an Appointment 安排會議

03 Request for Catalogues 索取目錄

04 Agenda for the Meeting 會議議程

05 Costumer Mailing List 顧客郵件地址清單

06 Conflict of Interest 利益衝突

07 Sales Policy 銷售政策

08 Sick Leave 病假請求

09 Vacation Request 休假請求

10 London International Business Show 倫敦商展

11 Cover Letter of Sales Manager 應徵業務經理一職

📍 日常生活類主旨書寫範例

01 RE: Lunch on Monday 回覆：週一午餐約會（主旨欄中時常出現的 RE：為 reply「回覆」的意思；而 FW：則為 forward「轉寄」之意。）

02 Can we meet Next Wednesday? 是否能於下週三會面？

03 We will arrive in Taiwan on Nov. 15th 我們將於十一月十五日抵達台灣

04 Hope you feel better now 祝您早日痊癒

05 Offer of Employment Notification 工作錄取通知

06 Letter of Recommendation 推薦函

07 Xmas Party 聖誕派對

6. Salutation 稱謂語

稱謂語通常位於信件主文的開頭，用來稱呼收件者。稱謂語後可接冒號 (:)（較為正式）或逗號 (,)（較為輕鬆）。如： Dear Ms. Wang: / Dear Ms. Wang,。

若是正式的書信往來，大多使用正式的招呼語 Dear 作為開頭，以表示尊敬。但現在電子郵件溝通有時為了創造輕鬆的氛圍，並適當拉近雙方的關係，有愈來愈多人偏向使用較不正式的 Hi, ... / Hello, ... 作為稱謂語， 看起來較為友善、有活力，和 Dear 相比，更接近現代電子郵件的寫法。

📍 稱謂語書寫範例

01 Dear Sir or Madam, 親愛的先生及女士（指稱公司全體）

02 Dear Sir, 親愛的先生（不知道男性的姓名時使用）

03 Dear Madam, 親愛的女士（不知道女性的姓名時使用）

04 Dear Mr. Wang, 親愛的王先生，（指稱已婚／未婚男性）

05 Dear Mrs. Lin, 親愛的林太太（指稱已婚女士）

06 Dear Miss Lin, 親愛的林小姐（指稱未婚女性）

07 Dear Ms. Lin, 親愛的林女士，（指稱已婚／未婚女性）

08 Ladies and Gentleman, 敬啟者（不確定指稱之對象及其性別）

09 To Whom It May Concern, 敬啟者（不確定指稱之對象）

10 Dear Product Managers, 親愛的產品管理人員（對象不只一人）

11 Hi, Everyone, 嗨，各位好（對象不只一人，且語氣較平易近人。）

12 Dear Lisa, 親愛的麗莎（指稱熟識的朋友）

※ 請特別注意 05 ～ 07 對女性的稱呼，若不確定收件者喜歡的稱呼為何，建議女性收件者以 Ms. 為主，以免因稱謂錯誤而冒犯對方。

7. Body of E-mail 信件主文（將於 Unit 2 詳細介紹）

8. Complimentary close 結尾敬語

撰寫英文 E-mail 時，通常會在信件的結尾留下問候語或祝福語，以表示禮貌。結尾後可接逗號 (,) 或不接，如 : Sincerely yours, / Sincerely yours。

📍 正式的結尾敬語

開頭指稱為 Dear Sir or Madam, Dear Managers, Ladies and Gentleman, 以及 To Whom It May Concern, 等信件時，建議搭配較為正式的結尾敬語：

01 Very truly yours,	**02** Truly yours,
03 Yours truly,	**04** Sincerely yours,
05 Yours sincerely,	**06** Faithfully yours,
07 Yours faithfully,	**08** Respectfully,
09 Respectfully yours,	**10** Yours respectfully,

📍 一般性結尾敬語

用於 Dear Mr. Chou, Dear Ms. Wang, Dear Mrs. Liu, 以及 Dear Miss Chou, 等信件時，可使用一般性的結尾敬語：

01 Sincerely yours,	**02** Sincerely,
03 Cordially yours,	**04** Cordially,
05 Thank you,	

📍 用於熟識友人之結尾敬語（開頭出現如 Hi Peggy, 等字樣）

01 Best regards,	**02** Best wishes,
03 Take care!	**04** All the best!
05 Continued success!	**06** Love,
07 With love,	**08** Cheers,
09 Good luck!	**10** Your devoted friend,
11 Warmest regards,	**12** Bye for now! = Till next time!
13 Have a nice/good/wonderful day!	

9. ~ 16. Signature and Information 署名與資訊

署名用以標示撰寫信件之書寫者，分為正式署名以及簡易署名兩種。正式署名涵蓋寄件人的完整資訊：

Part
1
入門篇

Part
2
求職篇

Part
3
客戶往來篇

Part
4
人際互動篇

Tina Yang → Signature 署名

Professor of Dept. of Foreign Languages and Literatures → Job title 職稱

National Taiwan University → Company name 公司名稱

No.1, Sec. 4, Roosevelt Rd., Da'an Dist., Taipei, Taiwan, R.O.C.
→ Company address 公司地址

Tel：(02) 2123-4567 → Company telephone number 公司電話

Fax：(02) 2123-4568 → Company fax number 公司傳真

E-mail：tinayang@ntu.edu.tw → Contact information 聯絡資訊

Website: www. front-page.com → Company website address 公司網站

然而，並不是每一封英文 E-mail 都必須包含完整的署名資訊。舉例來說，收件者可從寄件人欄位得知你的 E-mail，相關資訊就可省略，因此，在署名時應按照情況，選用簡易署名：

Grace Yang → Signature 署名

No.1, Sec. 4, Roosevelt Rd., Da'an Dist., Taipei, Taiwan, R.O.C.
→ Company address 公司地址

Tel：(02) 2123-4567 → Company telephone number 公司電話

Fax：(02) 2123-4568 → Company fax number 公司傳真

Tina Yang → Signature 署名

Professor of Dept. of Foreign Languages and Literatures → Job title 職稱

National Taiwan University → Company name 公司名稱

Grace Yang → Signature 署名

Unit 2 寫一封完美的 E-mail 主文
The Body of an E-mail

英文 E-mail 的主文（Body of E-mail）是信件的主體，所有需要讓收件者知道的訊息與指示，都會包含在主文中。

如同中文寫作一般，英文 E-mail 的撰寫也有起／承／轉／合；然而英文作文（尤其是商業書信）需要以簡短的文字點出重點，不求過於文情並茂，因此可以簡單分為三個段落：Introduction（寫信動機）、Development（要執行的動作）、Conclusion（結論）。

1. 寫 E-mail 常用的開頭句型（To begin a message）

1 We are writing to enquire about... / I'm reaching out to you to...
我來信的目的是要詢問有關⋯。／我來信的目的是為了⋯。

2 We are writing in connection with...
我來信是與⋯有關。

3 I am writing in reference to...
我寫這封信是關於⋯。

4 I am writing to you regarding...
我之所以寫信給您，是為了⋯。

5 After receiving your e-mail, I'm interested in...
收到您的來信後，我對⋯感興趣。

6 Thank you for your mail about..., I'm interested in...
謝謝您的來信，我對⋯感興趣。

7 And I would like to know...
而且，我想要知道⋯。

8 With reference to your e-mail of Jan. 20, I would like to know... / I would like to place an order for...
關於您於一月二十日的來信，我想知道⋯／我想下單購買⋯。

複製 / 貼上萬用句 Copy & Paste

We are writing to enquire about the price of your refrigerators.	我們來信詢問關於冰箱的價格。
We are writing in connection with your advertisement in the newspaper yesterday.	我們寫信來詢問關於昨天你們在報紙上刊登的廣告。
We are writing to request for the latest catalogue.	我們來信詢問您是否可以寄最新的目錄給我們。

2. 常用的回覆句型（To reply a message）

1 Thank you for your e-mail (of date) asking...
謝謝您（日期）的來信，詢問…。

2 Thank you for your fax enquiring about...
謝謝您的傳真，詢問…。

3 I have received your e-mail...
我收到您的電子郵件…。

複製 / 貼上萬用句 Copy & Paste

Thank you for your e-mail of July 11 enquiring about our prices.	謝謝您於七月十一日寄來的電子信件，詢問我們產品的價格。
Thank you for your e-mail asking if we would be interested in the new product.	感謝您的來信，詢問我們是否對新產品感興趣。
We have received your fax of August 26 concerning the exhibition in LA.	我們已收到您八月二十六日的傳真，關於在美國洛杉磯的展覽。
I received your e-mail today and I can't wait to thank you right away.	今天我收到了您的電子信件，並趕緊寫信向您致謝。

🖱	I'm so glad that you contact me again.	很高興您再度與我聯繫。
🖱	Thank you for your e-mail of last week. How have you been doing recently?	謝謝您上週寄來的電子郵件，您近來過得如何？
🖱	It is so delightful to have received your letter.	能收到你的來信真令人感到愉快。
🖱	Thank you for your e-mail/reply.	謝謝你的電子郵件／回覆。

📍 **回覆熟悉的對象（可於開頭加上簡單的問候，以示關心。）**

01 How have you been (doing) recently? 最近過得如何？

02 How are things? 近來可好？

03 How are you today? 今天覺得如何？

04 Are you doing well? 你過得好嗎？

05 How is the weather out there? 那裡的天氣如何？

06 Greetings from Taipei. 我從台北向您問候。

📍 **回覆對方問候的句子**

01 (I am) same as usual. / Same old, same old. 我和往常一樣。／老樣子。

02 Nothing special. / Nothing new. 沒什麼特別的。

03 Things are going well for me. 一切都很好。

04 Busy as always! 和往常一樣忙囉！

3. 常用的結尾句型（To end a message）

1 I look forward to... 期盼⋯。

2 I hope that... 希望⋯。

3 Please contact me if you... 如果你⋯，請與我聯絡。

4 Please let me know if you... 如果你⋯，請告知我。

5 Thank you for... 感謝您的⋯。

Part
1
入門篇

Part
2
求職篇

Part
3
客戶往來篇

Part
4
人際互動篇

複製 / 貼上萬用句 Copy & Paste

I look forward to hearing from you.	期盼收到你的來信。
I look forward to receiving your reply soon.	期盼盡快收到你的回信。
I look forward to getting your order.	期盼接到你的訂單。
I hope this information is helpful to you.	希望這個資訊對你有用。
Please contact me if you need any further information.	如果你需要更多資料,請與我連絡。
Please let me know if you have any problems.	如果你有任何問題,請告知我。
I will be in contact with you again.	我會再次與你聯繫。
We look forward to providing continued assistance and service.	我們期盼能與您長期合作。
Thank you again for your interest/ enquiry.	再次感謝您的關注 / 詢問。
Thank you for your cooperation.	謝謝您的合作。
Thank you for your continued support and loyal dedication to us.	謝謝您長久以來對我們的支持和生意往來。
(I will) Talk to you later.	以後再談。
Talk to you soon.	近期再聊。
Drop me a line soon.	有空寫封信給我吧!

4. 日期書寫方式(The date)

對於商業 E-mail 來說,日期往往涉及與客戶合作的細節(例如:出貨、繳款日期),所以很重要。在寫日期時,隨著美式 / 英式的不同,分為兩種寫法。

在英式英文中,日期會寫在前面;而美式英文則會將月份寫在前面。所以同樣的

10-8-2019，在英國就是八月十日，在美國則為十月八日，如此可能造成閱讀者的誤解，因此，最好的解法是將月份以英文寫出，日期則用數字表示，範例如下：

> **10 August, 2019**（英式寫法）或是 **October 8, 2019**（美式寫法）

📍 常見的日期呈現方式

01 Your order will be sent on June 12, 2020.

02 Your order will be sent on June 12th, 2020.

03 Your order will be sent on Tuesday, Jun. 12, 2020.

04 Your order will be sent on the 12th of June, 2020.

05 Your order will be sent on June of 12, 2020.

5. 縮寫式（Abbreviation）

除了正規的英文縮寫之外，網路「快速」的特性也發展出許多縮寫式，例如 Mr. 表示 Mister，BTW 表示 by the way。以下列出一些常用的縮寫式，但建議還是在通信雙方都非常了解英語用法的狀態下使用，如果不確定對方是否了解縮寫的意思，建議將全文打出，這樣除了能提升信件的專業度之外，也能避免造成收件人的困擾或誤解。

📍 月份（常用縮寫表示）

01 Jan. = January 一月

02 Feb. = February 二月

03 Mar. = March 三月

04 Apr. = April 四月

05 May 五月（無縮寫）

06 Jun. = June 六月

07 Jul. = July 七月

08 Aug. = August 八月

09 Sep. = September 九月

10 Oct. = October 十月

11 Nov. = November 十一月

12 Dec. = December 十二月

📍 星期

01 Mon. = Monday 星期一　　02 Tue. = Tuesday 星期二

03 Wed. = Wednesday 星期三　　04 Thu. = Thursday 星期四

05 Fri. = Friday 星期五　　06 Sat. = Saturday 星期六

07 Sun. = Sunday 星期日

📍 商用 E-mail 中的高頻縮寫字

01 BTW = by the way 順便一提　　02 PLS = please 請

03 THX / TKS / TNX = thanks 謝謝　　04 TMR = tomorrow 明天

05 AKA = as known as 也稱為是…　　06 ASAP = as soon as possible 盡快

07 TBD = to be discussed 有待商榷　　08 PIC = person in charge 負責的人

09 FYI = for your information 供您參考

10 TGIF = Thank god it's Friday. 終於星期五了！（慶祝週末的來臨）

📍 普通 & 商用 E-mail 皆可用的相關縮寫

01 COZ = because 因為　　02 CU = see you 再見

03 EZ = easy 簡單　　04 F2F = face to face 面對面

05 FAQ = frequently asked questions 常見問答集

06 B2B = business to business 公司對公司的商業模式

07 B2C = business to consumer 公司對顧客的商業模式

08 EOD = end of day/end of discussion 今天結束 / 討論結束

09 KPI = key performance indicators 關鍵績效指標

6. 表情符號（The Emoticon）

　　E-mail 不像面對面溝通，能看見雙方的肢體語言與表情，於是便衍生線條的表情符號，如最常見的 :-) 表示微笑，用以加強文字無法表達出來的情感。

　　不過，在商業 E-mail 裡使用表情符號必須小心，用得過多容易流於隨便，或給人不專業的印象；完全不使用又會過於嚴肅，難以產生親近感。因此，如何善用表情符號非常重要。以下介紹常用於 E-mail 的表情符號與建議的使用時機：

◉ :) 或 :-) 微笑　;) 或 ;-) 眨眼　:D 或 :-D 開心大笑　:-))) 非常開心

01 用於恭喜或祝賀的信件中，表達開心的情感。

02 用於感謝信函中，使語氣更加生動。

03 當內文涉及批評或指正時，可於糾正後另外加上微笑的符號，事先謝謝收件者願意接受指教，可緩和對方的情緒，提升溝通的順暢度。

04 開玩笑時，加上笑臉表情，可使收件者了解這只是個笑話，而不會當真。

◉ :-(難過、不開心、生氣　:,(哭泣　:-< 非常難過

01 傳達壞消息時加上表情符號，予人安慰之感。

02 指出對方的錯誤時，也可加上傷心的表情符號，表達希望對方能改進。

◉ 其他常見的表情符號

基本上可依心情自由挑選運用，但記得要適量，如果剛到新公司，也要注意辦公室氛圍是否適合使用表情符號；對於上司則勿隨便使用，以免觸犯上司。

01 B-) 戴著眼鏡微笑　　**02** :-o 表示訝異

03 :-P 吐舌開心的樣子　　**04** (@ @) 表示疑惑

05 :-O 哇！　　**06** :-| 無任何反應的表情

07 :-Q 抽菸　　**08** (^_^) 微笑

09 (^_*) 眨眼　　**10** (^o^) 哇！

11 (^*^) 親吻　　**12** (>0<) 糟糕！

13 (>.<) 生氣　　**14** (?_?) 疑問

15 (x_x) 疼痛　　**16** |-) 疲累

17 :-# 保守祕密　　**18** :-x 親吻；保守祕密

19 :/) 不有趣　　**20** (^^)// 鼓掌

Unit 3 商用 E-mail 的基本禮儀
Principles to Know before Writing

　　電子郵件風行，全台灣人也瘋上網，連中、小學都建置網路環境，教導小孩學習運用，再加上智慧型手機的普及，可以說，二十一世紀，將不會有「網路文盲」的存在。電子郵件，這絕不打烊的郵局，隨時隨地都在替全球的人們傳遞訊息。雖然隨著國家不同，寫信的風格會有所變化，但 E-mail 其實也有「世界通用」的標準化網際禮儀，一般需要注意的禮儀如下：

1 不可全用大寫（ALL CAPS），讀起來像對收件者吼叫，十分不禮貌。

2 態度有理，並適度使用感謝的詞句。

3 就算有摩擦，也要保持理性，並避免於情緒高漲時回覆信件。

4 電子信件的「主旨」需簡短並明確，讓人一目了然。

5 發送信件前，再檢查一次拼字、文法以及內容，勿直接發送初稿。

6 信件內容需簡明扼要，少用贅字，並確保主旨和內容一致。

7 英文修辭須減少被動語態，以避免句子冗長或模糊焦點等問題。

8 表情符號及現代英文的縮寫須謹慎使用，盡量少用於正式商業書信中。

9 若寄出時忘記附上附件，可再寄一封 E-mail，並於主旨標明：revised version with attachment。

10 回覆信件時須注意，只有在信件與所有收件者有關時，才使用「回覆所有人」（Reply All）的功能，否則寧可「回覆給寄件者」就好。

　　除了一般性的禮儀之外，也要注意 E-mail 內文的用字遣詞，以下列舉幾項常見的內容：

📍 不可不知的 E-mail 禮儀

01 說「請」或「謝謝」。　　　　　**02** 避免過於直接。

03 以「請求、拜託」代替「命令」。　**04** 使用婉轉語氣提出問題。

05 避免在信件中動怒或指責。　　　　**06** 負面的話要保守地說。

3-01 商用英文 E-mail 技巧：要求 vs. 請求

工作上，總會有意想不到的狀況需要別人幫忙；就算是沒有狀況時，也會需要和同事配合，共同執行一項工作。在與別人配合時，最直接的寫法是「下指令」，例如以下這封信：

From: Frank

To: Tim Watson

Subject: Sending E-mail to Mr. Crystal

Hi Tim,

Mr. Crystal from Minster Inc. called earlier and asked for our sales strategies for their new product. Please send our information to him at crystal@minster.com.hk by end of business today.

Frank

提姆好：

敏斯特公司的克里斯多先生剛剛打來，想知道我們打算如何行銷他們的新產品。請記得於今天下班前將相關資訊寄給克里斯多先生，他的信箱為：crystal@minster.com.hk。

法蘭克 上

如果我們是提姆，而法蘭克是你的上司，對方這樣寫也無可厚非。但若彼此是同事，不管對方是多麼重要的客戶，這種寫法都可能造成同事之間的誤會，反而無法順利完成工作。因此，除非特殊情況，否則請盡量避免這種「命令式」的寫法。

針對這封信，比較好一點的改法是「向對方解釋理由或情況，藉此增加對方配合的意願」，例如：

Part
1
入門篇

Part
2
求職篇

Part
3
客戶往來篇

Part
4
人際互動篇

From:	Frank
To:	Tim Watson
Subject:	Sending E-mail to Mr. Crystal

Hi Tim,

Mr. Crystal from Minster Inc. called earlier and asked for our sales strategies for their new product. **As I am leaving before noon for Hong Kong for business training,** please send our information to him at crystal@minster.com.hk by end of business today.

Frank

提姆好：

敏斯特公司的克里斯多先生剛剛打來，想知道我們打算如何行銷他們的新產品。因為我中午前就會離開公司，前往香港受訓，所以請你於今天下班前將相關資訊寄給克里斯多先生，他的信箱為：crystal@minster.com.hk。

法蘭克 上

　　同樣的內容，加上了理由（變色部份）就比較有禮貌，溝通的效果當然也會更好。但這種信件會讓收件者感受到「不得不」幫忙的壓力，因此建議不要常用。就這封信而言，信中所透露的「要求」多過「請求」。如果想讓彼此的溝通更加順暢，可藉由調整，將「要求」改為「請求」的語氣，下一頁將提供「請求」口氣的寫法。在閱讀時，會發現修改是從以下三點切入的：

1 表示需要對方幫忙，並說明情況。

2 要求最晚的回覆時間（by end of business today），也交代了原因。

3 內容完整，並加上委婉、感謝的字句（變色部分）。

From:	Frank
To:	Tim Watson
Subject:	Sending E-mail to Mr. Crystal

Hi Tim,

Need your help!

Mr. Crystal from Minster Inc. called earlier and asked for our sales strategies for their new product. **Unfortunately**, I am leaving for Hong Kong for business training before noon. **I'm wondering if you could** search your files and send our information to him at crystal@minster. com.hk.

If possible, please send him the information by end of business today as he needs it for a meeting with our product managers tomorrow.

I will appreciate your big help.

Frank

提姆好：

我需要你的幫忙！

敏斯特公司的克里斯多先生剛剛打來，想知道我們打算如何行銷他們的新產品。不巧的是，我中午前就必須離開公司，前往香港受訓，所以我想知道是否能請你查閱一下檔案，並將相關資訊寄給克里斯多先生？他的信箱為：crystal@minster.com.hk。

可以的話，請盡量於今天下班前將資料寄出，因為他明天會針對新產品，與我們的產品經理開會討論。

感謝你的幫忙。

法蘭克　上

3-02 禮多人不怪

在職場上，「禮貌與否」往往是合作能否順利進行的關鍵，以下列舉一些大家經常不自覺使用，但實際上給人壞印象的 NG 英語句。

📍 應避免的 NG 英文句

01 I got your e-mail. / I received your e-mail. 我收到你的信件了。

02 I haven't got your reply for three days! 已經三天沒有回音了！

03 Can you help me? 幫個忙吧？

04 I have to leave at 3 p.m., so come on time. 我必須在三點離開，所以請準時。

05 I'm too busy to join. 我太忙，無法參加。

06 What's your price for a XY25 printer? 你們型號 XY25 的印表機價格是多少？

07 Send me the price list. 寄價目表給我。

08 Remember to send me the quotation by next week. 下週前記得給我報價。

09 You must send it right now. 現在一定要寄出。

10 We cannot accept your prices. 我們不能接受你的價格。

11 There is a problem. 有個問題。

12 You've made a mistake with my order. 你把我們的訂單搞錯了。

13 I have sent the goods you ordered. You will get them soon. 已將你訂購的物品寄出，你很快就會收到。

14 Your company delivered the goods very late. This is a bad service. 你們公司的寄送速度其慢無比，服務真差。

15 Please deliver on time in the future. 以後請準時運送。

16 I can't help you on this issue. 這件事我不能幫你。

17 Please give us your valuable comments. 請提供寶貴意見。

上述句子的問題不外乎「命令式的語氣、過於直接、指責意味濃厚」這幾項。其實，無論狀況為何，在溝通時都應該保持基本的禮貌，才能順利讓對方配合處理問題。（針對上述的 NG 句，以下做了修改。）

✈ 複製 / 貼上萬用句 Copy & Paste

要求對方寄送資料

🖱 Can you please send me the price list?	可否請您寄價目表給我？
🖱 Can you (please) send it as soon as possible?	可否盡快寄出呢？
🖱 Would you send me the quotation by next week if possible?	可能的話，可以在下週前提供報價給我嗎？
🖱 We would appreciate receiving a quote of XY325 printer, including price list, tax, packing, and delivery fee. Thank you.	可否告知型號 XY325 印表機的價格？若能給我包含稅金、包裝以及貨運的完整價格，我將非常感激。

各式各樣的問題

🖱 I was wondering if you could help me.	不知道您是否能幫我一個忙？
🖱 We welcome any comments you may have.	我們誠摯地邀請您提供意見。
🖱 May I have some update about this issue?	請問這件事情有進展嗎？
🖱 We think your prices are rather high.	我們認為貴公司的價格有點高。
🖱 I am afraid there is a problem with the order.	很遺憾，訂單出了一點狀況。
🖱 There seems to be a small problem.	似乎有個小問題。
🖱 We have received Order789, but it was delayed.	我們收到了 789 號訂單的貨品，但遺憾的是，它到得有點晚。

We would be grateful if our orders can arrive on time in the future.

若未來的訂購商品能準時送達的話，我們將非常感激。

If you're available at 11:00 next Thursday, I will make a reservation at ABC Cafe.

若您下週四上午十一點方便的話，我會預先訂好 ABC 咖啡廳的座位。

My next appointment is at 15:00, so it'll be great if you can come on time.

因為我三點另外有約，必須先離開，所以若您能準時抵達的話，我會非常感激。

回覆信件

Thank you for your e-mail.

謝謝您的來信。

Thank you for your e-mail. The goods had already been sent to you, and should be arrived by Wednesday.

感謝您的來信，貨品已寄出，您應該會於週三前收到貨品。

Unfortunately, I'm in the middle of something.

不巧我剛好有事。

Sorry, but that isn't my strong suit.

不好意思，這不是我的強項。

實力大補帖 Let's learn more!

深入學 資訊深度追蹤

📍 E-mail 內文的注意事項

　　商務 E-mail 最重要的目的，在於溝通。即便在需要嚴厲表態的時候，也得避開情緒性的字眼，才能讓對方冷靜地與你溝通。以下再提醒大家兩點：

01 使用專業的稱呼：稱呼與招呼語能展現專業度，千萬不要讓一個稱呼，讓對方認為你態度輕率。

02 不要濫用驚嘆號：過多的驚嘆號只會讓人覺得你情緒不穩，因此須避免。

筆記頁

Part **2**
求職篇

職場新鮮人必備

@ 本章焦點 Focus！

✔ 履歷 & 自傳	✔ 有力的推薦信	✔ 爭取面試機會

　　社會新鮮人所要踏出的第一步，就是求職。如何在千萬封履歷中脫穎而出，就要靠吸睛的履歷 & 求職信了。而就公司方面而言，就會有邀請應徵者來面試，以及通知結果等 E-mail 必須撰寫。與求職有關的內容，看這一章就足夠。

打造完美履歷表
Writing Your Resume

1-01 履歷表

　　履歷表的格式和內容並沒有固定規則，但基本資料是不可或缺的，例如：姓名、地址、電話號碼、工作經驗、教育程度、技能及興趣等。要注意的是，無論是什麼樣的履歷表，都必須扼要地展示出個人生活的各種面向，以幫助雇主認識應試者。

　　英文履歷主要分為兩種：Resume 和 Curriculum Vitae（CV）。Resume 是簡單扼要的個人背景概述，長度最多不超過 2 頁，較常見於一般求職；CV 則是較詳細的履歷，內容包含學術經驗、出版刊物、研究經驗等，較常用於申請學術相關的單位。

　　除了履歷表之外，應徵外商公司通常還須附上求職信（Cover Letter），針對某特定職缺表明興趣，並針對該職缺，詳述相關資格與能力，簡述自己能勝任的原因。

　　無論應徵公司為何，履歷都是公司評估應徵者的主要文件，也是考量是否要請對方來面試的審查依據。因此，本單元將會提供履歷表與自傳範例，請大家在實際撰寫時，依公司的要求和個人情況做調整。

1. 履歷表的段落標題

　　現代社會的競爭激烈，求職者在選擇屬意的公司時，其實也正面臨被雇主挑選的命運。這種「雙向性」的互動導致應徵者必須扮演「業務的角色」，將自己推銷給雇主，而自我推銷的媒介，就是投遞的這份履歷表。它等同於業務推銷產品的資料，使雇主能迅速決定「買」或「不買」。

　　一般而言，履歷表所涵蓋的項目無外乎下面幾項：

1 Objective 求職目的

2 Desired Position / Position Applied For 應徵職位

3 Educational Background 教育背景

4 Related Courses 相關課程

5 Employment History / Work History 工作經歷

6 Related Experience 相關經驗

7 Relevant Skills 相關技能

8 Qualifications 檢定考試

9 Honors And Activities 社團與獲獎事蹟

10 References 推薦者

11 References are available upon request. 推薦函備索（通常於底部置中）

12 Degree expected June 2019 預計於二〇一九年六月畢業（用於在校生）

　　雖然上面列舉了十二項的段落標題，但不同職位或性質的工作其實會對應徵者有不同的要求，因此，撰寫履歷時，應該以對方公司的需求為優先考量，來決定該突顯自己哪方面的特質。

　　總之，一份好的履歷，必須有所取捨，才能成為你求職面試時的促銷資料。

2. 打造一份成功履歷的必要關鍵

　　理解了履歷表所涵蓋的內容，並不表示就能寫出一份打動雇主的履歷，如果沒掌握撰寫時的關鍵，很可能寫出平凡無奇、毫不吸睛的內容，為了幫助大家避開撰寫時的地雷，以下特別提供撰寫履歷時一定要掌握的要項。

1 短小精簡，直擊核心：要讓雇主能在短短一瞥的幾秒內，馬上判斷「你的能力是否符合需求」。因此，履歷表的頁數最好為 1-2 頁，盡量在這個頁數範圍內展示自己的長處，提供與應徵職位相關的資訊。

2 先重後輕：一般人習慣從頭到尾按順序閱讀，尤其是對快速審閱的雇主來說，先接觸到的資訊往往比後接觸到的印象深刻。因此宜將「最有利、與所應徵職位最相關的資訊放在前面」。

3 學歷就高不就低：同上點，學歷部分也應該從最高學歷寫起，依序往前推，使雇主能對求職者的教育程度一目了然。另外，除非公司特別要求，否則國小及國中等學歷可省略。

4 表達宜簡潔：建議以「短句條列式」說明相關經歷。除了與應徵職位相關的專業技能可使用專業術語，其他地方宜採用淺顯易懂的文字，避免使人難以閱讀（或者產生誤會）。

5 列出與應徵職位相關的技能與經驗：履歷中的所有內容均須利於自己爭取到該職位。但要注意的是，在敘述能力時，宜用客觀的事實呈現，避免誇示。

6 提供清楚的聯絡方式。

General Resume 1

Pin-Tseng Liu

109 LongChian St., Taipei 105

(02) 7788-9999 * pintsengliu@mail.net

EDUCATION

UCLA Anderson, School of Management, Los Angeles, California, U.S.A.

Master of Business Administration, Expected June 2018

■ Member of International Business Union (2017)

■ Graduate Business Club (2016 - present)

National Chung Hsing University, Taichung, Taiwan

B.A. in Marketing, Minor in History (June 2012)

■ Outstanding Scholarship

■ Exchange Student in the University of Eastern London

EXPERIENCE

Johnson & Mary Inc., Taipei, Taiwan

Investment Banking Division, Analyst (2013 - 2015)

■ Provided investment banking services to private investment firms.

■ Advised buyers on financing concerns.

■ Led the development of financial models in the company.

Black-White Corp., Taichung, Taiwan

Marketing Division, Officer (2012 - 2013)

■ Organized marketing groups, developed departmental marketing procedures and strategies, and delivered presentations.

SKILLS

Computer

■ MS Excel, Access, Word, PowerPoint

Language

■ Proficient in spoken and written English

■ Conversational in Japanese

一般履歷表（一）

劉品貞

105 台北市龍江街 109 號

(02) 7788-9999 * pintsengliu@mail.net

教育背景：

美國加州大學洛杉磯分校，安德森管理學院

工商管理碩士，預計 2018 年六月畢業。

■ 國際商業組織協會會員（2017 年）

■ 碩士生商業社團（2016 年至今）

國立中興大學（台中市）

主修：行銷；副修：歷史。2012 年六月畢業。

■ 獲得優秀學生獎學金

■ 東倫敦大學交換學生

工作經驗：

台北市強生 & 瑪麗公司，銀行投資處分析師（工作期間：2013 年至 2015 年）

■ 提供私人投資公司銀行投資的服務。

■ 提供投資者財務諮詢的服務。

■ 協助規劃公司內部的財務發展。

台中市黑白公司，行銷部專員（工作期間：2012 年至 2013 年）

■ 組織行銷團隊，規劃部門行銷程序及策略發展，做行銷簡報。

技能：

電腦

■ 微軟試算表、微軟資料庫管理系統、微軟文書處理及簡報軟體

語言

■ 精通英語，說寫流利

■ 基本日語溝通能力

Pei-Fen Yang

1380 HeiKu E. Ave.

New Taipei City

0911-555-666

EDUCATION

Taipei Community College, December 2015

Certificate in Nursing

RELATED COURSES

Introduction to Nursing, Nursing Seminar, Psychology, Developmental Psychology, Sociology, Child Health Nursing

RELATED EXPERIENCE

Hospital Volunteer

Chang Gung Memorial Hospital (March 2013 - August 2016)

■ Assisted in the emergency room.

First-Aid Course Assistant

ABC Girl Scout Council (December 2011 - December 2012)

■ Assisted instructors in Girl Scout in first-aid courses.

WORK EXPERIENCE

Full-time Nanny

New Taipei City (September 2014 to present)

Part-time Cashier

7-11 Convenience Store, New Taipei City (March 2012 to March 2013)

Part-time Receptionist

MOS Burger, New Taipei City (July 2008 to June 2010)

HONORS AND ACTIVITIES

■ Outstanding Mother, New Taipei City, 2016

■ Girl Scout Troop Leader, New Taipei City Junior High School, 2009 - 2011

一般履歷表（二）

楊佩芬

新北市黑酷東大道 1380 號

0911-555-666

教育背景

台北社區大學，護理證照。2015 年十二月畢業。

相關修業課程：

護理入門、護理研討、心理學、發展心理學、社會學、兒童健康護理

相關經驗：

醫護志工

長庚紀念醫院（2013 年三月至 2016 年八月）

■急診室醫護協助。

急救課程助理

ABC 女子童軍協會（2011 年十二月至 2012 年十二月）

■協助女童軍講師教授急救課程。

工作經驗：

全職保母　　新北市（2014 年九月至今）

兼職收銀員　7-11 便利商店，新北市（2012 年三月至 2013 年三月）

兼職接待員　摩斯漢堡，新北市（2008 年七月至 2010 年六月）

獲獎事蹟

■2016 年新北市傑出母親

■2009 年到 2011 年新北市國中女童軍團團長

Position with an Emphasis on Child Development

Maggie Wang

Address Until June 2019

National ABC University

P.O. Box 30123

Taipei 105

(02) 3456-7890

maggiewang@mail.net

Permanent Address

999 ChungShan Road

WuFeng, Taichung

(04) 1234-5678

OBJECTIVE   To obtain a position with an emphasis on child development in which I can contribute my caring skills.

EDUCATION   National ABC University, Taipei

Bachelor of Arts in French, expected June 2019

Cumulative GPA: A / 85

XYZ Senior High School, Taichung (September 2013 to June 2016)

Academic Honors:

Cathay Scholarship, First Care Award

Related Courses:

Child Development, Adolescence and Youth, Social Psychology, The Family Phenomenon

EXPERIENCE   *ABC Children's Center, Taipei. Caretaker. (Fall 2016)*

- Supervised 5 - 10 children (infant - 2 years) in mealtime, nap set-up, and playtime.

- Promoted social interaction among children.

- Comforted upset and distressed children.

ABC Office of Kid's Life, Taichung. Adviser (Summer 2013)

- Planned monthly programs for kids (3 - 6years old).

- Offered emotional support and kid counseling.

- Demonstrated skills for handling emergency.

兒童發展相關工作

王梅姬

居住地址（至 2019 年六月）：

國立 ABC 大學

105 台北市

郵政信箱 30123 號

電話號碼：(02) 3456-7890

maggiewang@mail.net

永久地址：

台中市霧峰區中山路 999 號

電話號碼：(04) 1234-5678

求職目標：能貢獻所學之兒童發展相關工作

學歷：台北國立 ABC 大學法文系

預計 2019 年六月畢業

學期成績：A / 85 分

台中 XYZ 高級中學

求學期間：2013 年九月至 2016 年六月

學業表現：

國泰集團獎學金、第一保育協會獎學金

相關課程：

兒童發展課程、青少年和幼兒課程、社會心理學、家庭現象課程

經驗：台北市的 ABC 幼兒中心，照顧者。（2016 年，秋）

- 管理五至十名幼童（嬰兒到兩歲幼兒），於用餐時間、午休、下課時間給予指導和協助。
- 促進孩童間的互動。
- 安撫情緒不穩定的孩童。

台中市的 ABC 幼兒生活館，顧問。（2013 年，夏）

- 規劃孩童（三到六歲）每月的活動課程。
- 提供情緒上的協助。
- 具備處理緊急狀況的專業能力。

Human Resources Representative

Danny Chou

123 Ocean Street

Onetown, ShinChu

(03) 8888-9999

OBJECTIVE

To obtain a human resources representative position in which I can contribute my headhunting skills.

SUMMARY

- Human resources internship, with XL Company.

- Proficient in MS Office, Windows, and IE (Internet Explorer)

EDUCATION

XYL Technology University, Tainan

Bachelor of Business Administration, June 2018

- GPA: 83 / 100

- Courses on Human Resources Management, Management Theory, Corporate Communications, Human Relations

EXPERIENCE

XL Company, Tainan

Human Resources Intern, Jun. to Aug., 2017

- Supported recruiting

- Responsible for developing position requisitions

- Posted job vacancies and conducted interviews for candidates

First Bank, Tainan

Office Assistant, summer 2015 and 2016

- Processed monthly statements

- Developed system for employee file documentation

ACTIVITIES

- Member, Association of Human Resources, XYL Technology University, 2016 - 2017

- Member of tennis team, 2015 - 2018

人力資源部門代表

周丹尼

新竹市第一鄉海洋街 123 號

(03) 8888-9999

求職目標：在人力資源部門發揮所長。

摘要：
- 曾任 XL 公司人力資源部門實習生。
- 熟悉微軟辦公室軟體、微軟系統、網際網路。

教育背景：

台南 XYL 科技大學

商業管理學系，2018 年六月畢
- 畢業成績：83 分
- 修習課程包含：人資管理、管理學理論、公司團體溝通法、人際關係

工作經驗：

台南 XL 公司，人力資源部門實習生（2017 年六月至八月）
- 協助公司招募員工
- 負責職缺的申請過程
- 刊登職缺、協助面試

台南第一銀行，辦公室助理（2015 年及 2016 年的夏季）
- 負責處理每月銀行結單
- 開發職員檔案文件系統

參與活動：
- XYL 科技大學人力資源協會會員（2016 年至 2017 年）
- 網球隊隊員（2015 年至 2018 年）

A Position in Electronic Engineering

Bill Lin

12F, No.77 Sanmin Road, Kaohsiung

(07) 1234-6666

billlin@mail.net

OBJECTIVE

To obtain a position in electronic engineering in which I can utilize my professional skills in engineering.

EDUCATION

B.S. in electronic engineering, University of Michigan, May 2015

- GPA: 3.5 / 5
- Courses include: Design and Testing of Auto Structure, Advanced Dynamics, Linear Systems Analysis

EXPERIENCE

Engineer, GL Automobile Company, Detroit

July 2016 - July 2018

- Assisted with testing and analysis for V36 vehicle and revised computer programs.

Engineering Assistant, GL Automobile Company, Detroit

January - June, 2016

- Assisted mechanics in repairing automobiles and dealt with customer issues.

COMPUTER SKILLS

CAD	AutoCAD	C, C++
MS Word	MS Excel	Adobe Photoshop

HONORS AND ACTIVITIES

- Auto Prize in Electronic Engineering (Spring 2017)
- TA for Calculus I and II (Fall 2014)

References available upon request.

林比爾

高雄市三民路 77 號 12 樓

電話：(07) 1234-6666

電子郵件：billlin@mail.net

求職目標：

應徵電子工程職缺，以發揮所學。

教育背景：

美國密西根大學，電子工程學系。2015 年五月畢業。

■ 畢業平均成績：3.5 / 5

■ 修習課程包含：汽車結構設計及測試、進階動力學、線性系統分析

工作經驗：

底特律 GL 汽車公司，工程師（2016 年七月至 2018 年七月）

■ 協助 V36 車型的測試與分析、更新電腦程式。

底特律 GL 汽車公司，工程助理（2016 年一月至六月）

■ 協助人員維修汽車、處理顧客的問題。

電腦技能：

■ 電腦輔助設計（Computer Aided Design, CAD）

■ 美國 Autodesk 公司之電腦繪圖軟體（AutoCAD）

■ C 語言、C++ 語言

■ 微軟文書處理軟體、微軟試算表、Adobe 影像處理軟體

社團與獲獎事蹟：

■ 汽車電子工程獎（2017 年，春）

■ 微積分 I 與微積分 II 助教（2014 年，秋）

推薦函備索

A Position in Software Development

//

Peter Wang

Current Address	**Permanent Address**	**E-mail Address and URL**
88 Tiger Street	45 Gates Way	peterwang@mail.net
Hualien	Hualien	http://www.hualien.edu/peterw

OBJECTIVE

To obtain a position in software development in which I can contribute my professional skills.

EDUCATION

B.S. in Computer Science (June 2014)

National Dong Hwa University

■ Cumulative GPA: 85 / 100

WORK EXPERIENCE

Software Computer Consultant (September 2017 to present)

■ Create World Wide Web home pages and customize computer systems for clients at *Orange Computer Inc., Hualien.*

Office Assistant (Summer 2015 to 2016)

■ Maintained customer database, answered phones and replied correspondence at *National Dong Hwa University.*

COMPUTER SKILLS

Languages

C, C++, Java, Java Script, UNIX, HTML

Operating Systems

UNIX, Windows, Macintosh

Software

MS Visual C++, MS Word, MS Access, MS Excel, Adobe Photoshop, Adobe Acrobat

Hardware

HP-UX, PC, Macintosh

王彼得

居住地：
花蓮縣老虎街 88 號

永久地址：
花蓮縣大門街 45 號

電子郵件和網站：
peterwang@mail.net
http://www.hualien.edu/peterw

求職目標：應徵軟體開發職缺，以發揮所學與專業。

教育背景：

國立東華大學 資訊工程學系（2014 年六月畢業）

■ 畢業成績：85 分

工作經驗：

花蓮市橘子電腦公司，**電腦軟體顧問**（2017 年九月至今）

■ 設立網頁和客戶電腦系統。

國立東華大學，**辦公室助理**（2015 年夏至 2016 年）

■ 維護顧客資料，回覆電話及電子郵件。

電腦技能：

電腦語言程式

C 語言、C＋＋語言、Java 語言、Java Script 語言、UNIX 操作系統、HTML 網頁語法

操作系統

UNIX 操作系統、Windows 微軟作業系統、Macintosh 麥金塔作業系統

軟體

微軟 C++ 語言、微軟文書處理軟體、微軟資料庫管理系統、微軟試算表、Adobe 影像處理軟體、Adobe 文件處理器

硬體

HP-UX 電腦、個人電腦、蘋果電腦

複製 / 貼上萬用句 Copy & Paste

說明工作經驗

Taught ESL (English as a Second Language) to small groups of children and adults.	曾教授小班制兒童與成人的英語課程。
Provided administrative support for schools.	曾為學校管理提供行政方面的支援。
Scheduled conferences and seminars for senior executives.	安排所有高階主管的會議與研討會事宜。
Executed marketing and advertising plans for an E-book manufacturer.	曾為電子書產品的製造商製作行銷與廣告企劃。
Delivered presentations to international clients to expand the market.	為國際客戶進行簡報，以擴展市場。
Increased sales by 35% through innovative sales skills.	曾透過創新的銷售技巧提高了百分之三十五的業績。
Responsible for special orders and customer relations.	負責處理特殊訂單與維護顧客關係。
Handled all sales transactions in men's and boys' wear.	管理所有男性服飾的交易。
Responsible for purchasing all men's clothing and accessories.	負責所有男性服飾和配件的採購事宜。
Handled budgets up to $250,000.	管理高達二十五萬美元的預算。
Increased sales by 25 percent in two years.	在兩年內增加百分之二十五的銷售量。
Supervised three assistants and one secretary.	管理過三位助理及一位祕書。
Handled general secretarial duties in business administration department.	處理商業行政科一般性的祕書事務。

強調技能和興趣

Computer skills: MS Office, Photoshop, PageMaker, and Dreamweaver	電腦技能：微軟辦公室軟體、繪圖、排版與網頁製作軟體
Language skills: intermediate level of Japanese and Korean	語言能力：日文與韓文中級程度
Skills: shorthand 100 wpm, English typing 60 wpm	技能：速記每分鐘一百字，英打每分鐘六十字
Qualifications: New TOEIC Test Total Score 860; Listening: 460; Reading: 400	檢定考試：新多益測驗總成績 860 分；聽力 460 分；閱讀 400 分
GEPT Certificate of General English Proficiency – Intermediate	全民英檢中級證書
IELTS Certificate (International English Language Testing System): Total Score 7	雅思檢定測驗：總分 7 分
Other activities: member of the football team, member of debate society of the Taiwan University	其他活動：台大足球隊隊員、辯論社社員
Hobbies: cooking, swimming, singing, writing, photography	嗜好：烹飪、游泳、唱歌、寫作、攝影

實力大補帖 Let's learn more!

活字典 單字 / 片語集中站

英文縮寫與全稱

Apt. = Apartment 公寓	**Ave. = Avenue** 大街；大道
Bldg. = Building 大樓	**Blvd. = Boulevard** 大道

St. = Street 街道;路	Co. = Company 公司
Corp. = Corporation 公司	Inc. = Incorporated(有限)公司
Dept. = Department 部門	Mfg. = Manufacturing 製造商
E.= East 東	W. = West 西
S. = South 南	N. = North 北

常見的職業名稱

accountant 會計師	actor / actress 男 / 女演員
anchor 新聞主播	architect 建築師
artist 藝術家	associate professor 副教授
astronaut 太空人	auditor 審計員;稽核員
baker 烘焙師	blacksmith 鐵匠
broker (agent) 經紀人	bus driver 公車司機
carpenter 木匠	cashier 出納員
chef / cook 廚師	chemist 化學家
clerk 店員	customs officer 海關官員
dentist 牙科醫生	designer 設計師
doctor 醫生	diplomat 外交官
editor 編輯	electrician 電工
engineer 工程師	fashion designer 時裝設計師
general manager 總經理	guard 警衛
housekeeper 管家	interpreter 口譯員
journalist 記者	janitor 守衛;管理員
lawyer 律師	librarian 圖書館員
mechanic 技工	model 模特兒
musician 音樂家	nurse 護士
pharmacist 藥劑師	photographer 攝影師

assistant 助理	receptionist 接待員
secretary 祕書	soldier 士兵；軍人
technician 技術人員	tour guide 導遊
translator 筆譯；譯者	TV producer 電視製作人
vet / veterinarian 獸醫	writer 作家

📍 較專業的職業名稱

announcer 廣播員	archaeologist 考古學家
actuary （保險）精算師	baseball player 棒球選手
boxer 拳擊手	butcher 屠夫；肉販
cartoonist 漫畫家	clown 小丑
cobbler 製鞋匠；補鞋匠	counselor 顧問
civil servant 公務員	dancer 舞者
detective 偵探	farmer 農夫
firefighter 消防員	fisherman 漁夫
florist 花商；花藝師	foreign minister 外交部長
gardener 園丁	geologist 地質學家
hairdresser 美髮師	judge 法官
life guard 救生員	illustrator 插畫家
magician 魔術師	mathematician 數學家
miner 礦工	movie director 電影導演
nanny 褓姆	policeman 警員
postal clerk 郵政人員	postman 郵差
priest 牧師	professor 教授
sailor 水手	scientist 科學家
tailor 裁縫師	trainee / intern 實習生

寫一封吸引人的自傳
A Perfect Autobiography

2-01 範例：應徵工作的自傳

在寫求職用的自傳時，身家背景建議以少量篇幅帶過，盡量著重在與應徵職位相關的經歷，以吸引面試官的注意。

I was born in Beitou, Taipei, Taiwan on March 19, 1989 of a middle class. My father is a product manager at the Department of Marketing, XYZ Company. My mother is an English teacher in an university.

Although I am an only child, I am by no means a spoonfed girl. I was strictly taught to behave when little, and my early school experiences taught me to get along with people. In fact, the friendship among my schoolmates and me still lasts until now.

School life in the capital city, Taipei, is extremely competitive, but I managed to keep my academic records above average all the way from primary school to university. Not only did I keep good control of my academic performance, I also enjoyed school life and joined many extracurricular activities, such as the International Affairs Seminar, Youth Study Tour to Singapore which was sponsored by Taipei City Government. Last summer vacation, I was employed as an assistant Chinese language teacher in the Chinese language and culture program in Taiwan, sponsored by California University. Besides, I have been a tutor all these four years, and I really become interested in teaching.

I have been trying to fulfill my life by trying different things and having various experiences. I hope my experiences can bring benefits to your institution.

一九八九年三月十九日那一天，我出生於台北市北投的中等家庭。父親是 XYZ 公司的產品經理，母親為大學的英文教師。

雖然我是獨生女，卻沒有因此而嬌生慣養。從小時候開始，母親就對我的言行舉止很注意，並加以管教，而我的小學生活也教會我與人友好相處。事實上，我和那時候的同學們到現在都還有聯絡。

雖然台北人的課業競爭十分激烈，但從小學到大學，我的在校成績始終保持在平均分數以上。我的在校生活過得很開心，除了課業之外，我還參加了許多課外活動，例如

國際事務營、台北市政府舉辦的新加坡青年留學計畫……。去年暑假我受聘為助教,在加州大學於台灣開辦的中文及文化課程教中文。除此之外,在大學四年期間,我一直擔任家教,對教書工作充滿濃厚的興趣。

我一直在找機會充實自己,嘗試各種不同的事,並藉此增加自己的經驗。期盼我能將自己的種種經歷,轉化為對公司的貢獻,替公司帶來利益。

✈ 複製 / 貼上萬用句 Copy & Paste

自我介紹與家庭背景

My name is Chun Yu Lee. I was born in 1989, and had lived with my family in Taichung for twelve years before we moved to Taipei ten years ago.	我叫李均昱,出生於一九八九年,在台中住了十二年,十年前舉家遷居台北。
There are three boys in my family. I am the second son.	家中有三兄弟,我排行老二。
My name is Sing Hua Hu. My father came from Hong Kong but I was born in Taiwan in 1990.	我叫胡欣樺,父親本籍香港,而我於一九九〇年出生於台灣。
I am the youngest daughter but no one can find in me any trace of a spoiled child.	雖然我是最小的女兒,但卻沒有因此被寵壞。
My mother passed away when I was six and my father had to take on the responsibility of taking care of the family.	母親不幸在我六歲時過世,所以父親肩負起照顧我們的責任。
Instead of being self-indulgent, I grew up with an outgoing and open-minded personality.	我的個性外向,喜愛體驗新奇事物,一點都不驕縱任性。
My parents moved to Taiwan in 1985 when I was 5 years old.	父母親在一九八五年(也就是我五歲時)移居台灣。

Because my parents grew up without much education and worked laboriously for a living, they wanted me to receive the best education in Taipei.

因為雙親從小沒唸多少書，必須努力工作才得以糊口，所以希望我在台北受最好的教育。

求學過程

When I was a child, I was very interested in arithmetic. I score higher in mathematics than other subjects, and I'd love to be an accountant.

我從小就對算術極感興趣。求學期間，我的數學成績總是超過其他科目，因此我想當一名會計師。

I have always had a burning desire for chemistry, and I hope to be an expert in applied chemistry. Therefore, I took lots of chemistry related courses and did extra reading and laboratory work.

我一直對化學有濃厚的興趣，希望有一天能成為應用化學的專家。因此大學選修了不少學分，也投注許多心力閱讀書籍及做實驗。

During summer vacation, I worked at the laboratory of Chung Shan Pharmaceutics Co. Ltd. to earn tuition fees and most important of all, to obtain experiences and knowledge in this field.

暑假期間，我去中山醫藥公司的實驗室上班，以賺取學費，但比錢更重要的是從中獲取實務經驗與相關知識。

I have great interests in computer. I have studied computer for almost seven years. My computer skills are superb. My friends always come to me for their computer problems.

我對電腦有濃厚的興趣，我研究電腦將近七年，電腦能力一流，所以朋友總是來問我一些有關電腦的問題。

I know how to set up a website and make flash animations. Currently, I have my own website sharing my habits and interests with people around the world.

我知道如何架設網站和製作動畫。目前我有個人網頁，藉此與世界各地的人們分享我的嗜好和興趣。

未來展望

Having completed the military service, I entered a trading corporation as an assistant, dealing with machinery import.	服完兵役後，我進入一家貿易公司擔任助理，負責機械的進口事宜。
Now I hope to work for an American corporation so that I can be more independent.	我希望到美商公司工作，使自己更為獨立自主。
Having obtained practical experiences in accounting by working as an accountant at XYZ Company, now I consider it's time to move forward to a larger corporation.	我曾於 XYZ 公司擔任會計，因此獲得實務經驗，而現在，我認為自己已經準備好往大公司發展了。
Being cautious and conscientious, I am sure that I can take greater responsibility at work.	我認為自己做事仔細、具責任感，相信我可以擔負起更大的責任。
With my computer and English abilities, I am confident to excel in this field.	基於我所具備的電腦和英語能力，我有自信能在這個領域交出一份漂亮的成績。
I am eager to be a programmer who develops information technology to contribute to modern technology in the world.	我期盼能成為一名發展資訊科技的電腦工程師，為這個社會盡一份心力。
I hope that you can offer me the opportunity to prove myself.	希望貴公司能給我一個證明我能力的機會。

2-02 各領域的自傳要點

應徵各領域的工作時，皆應表明想進入該領域的動機或緣由，及自己做過的研究和努力，但隨著個人所屬學系的不同，自傳中可特別強調一些要點，以提升被錄取的機率。以下就針對常見的幾個學系做介紹：

1. 財經學系

1 強調自己的外語、數理、電腦操作等能力

2 擔任幹部或領導者的經歷

3 團隊合作經驗，以及在團隊中所扮演的角色

2. 法政學系

1 演講比賽的經驗　　　　2 曾辯論的議題與結果

3 詳細說明個人特質（好惡分明、實事求是等）

4 強調自己的學習態度或方法，並闡述對自己的期許和規劃

3. 醫藥衛生學系

1 說明人格特質，強調責任感　　2 舉例說明關懷他人之心

3 人際關係與溝通能力　　　　　4 對於研究、查資料的概念

5 各項外語能力

4. 管理學系

1 對數字的概念　　　　　2 對市場競爭力的想法

3 如何評估事情的成功與否　4 自我「組織管理」的方法

5 是否觀察過公司經營型態或市場活動的經驗

5. 資訊學系

1 邏輯思考的能力　　　　2 吸收新知的方法

3 動手操作的經驗　　　　4 對數學的概念

6. 工程學系

1 對物理、微積分、化學的喜好

2 動手製作實物的喜好及經驗

3 解決問題時的毅力與堅持（舉例說明為佳）

4 女同學可突顯對於設計、研發、品管方面的興趣

7. 社會心理學系

1 強調社團活動經驗　　**2** 觀察人類行為的心得

3 是否擔任過義工或志工　　**4** 提出對社會議題的想法

5 具體說明遇到挫折或困難的處理方式

8. 教育學系

1 自己適合當老師的原因　　**2** 社團經驗

3 舉例說明自己的相關能力（教學、輔導、溝通等）

4 舉例說明自己具備的毅力與耐心

5 影響自己最深的人或老師

9. 外語學系

1 最喜歡的作家與書籍　　**2** 增進語文能力的方法

3 參加校內外活動或競賽的經歷

4 若所學為非英語系，可說明為何選擇該語言及對該國的認識

10. 大眾傳播學系

1 平日接觸媒體的習慣　　**2** 對人事物的好奇心

3 遇到問題的思考模式　　**4** 隨機應變的能力與經驗

5 對於媒體的看法　　**6** 語文能力、行動力、溝通能力

11. 藝術學系

1 對藝術的敏銳度　　**2** 對藝術的感受力

3 意志力與韌性　　**4** 勇於接受挑戰

5 獨樹一幟的創造力　　**6** 提供作品集

用推薦信替自己加分
The Recommendation Letter

3-01 請求協助寫推薦信

　　找工作時，有時雇主會要求提供前上司的推薦信函，而一封專業的推薦函確實也能為履歷加分不少，因此，以下將從「請求寫推薦信」開始逐一介紹。

　　若要麻煩他人為你寫推薦信，記得要把下列內容寫入信中：一、想應徵的職務；二、請對方當推薦人的理由；三、告知應聘公司的信箱及聯絡人資訊，並感謝對方協助。

From:	Laura Chen
To:	Prof. Lin
Subject:	Request for Job Reference

Dear Professor Lin,

I am in the **process**[1] of **seeking**[2] a new position as a sales assistant in Costco Wholesale **Corporation**[3] in Chicago, and am hoping that you would provide a reference for me.

Having worked for you for many years, I'm sure that you know my **attitude**[4] at work, as well as my **strengths**[5] and skills better than anyone. It would be helpful if you could provide my **potential**[6] employer information about my strengths and skills to increase the chances for me to get the opportunity.

I have enclosed a copy of my resume. Please contact Ms. Green, Vice President at Costco. And please let me know if there's any additional information you may need.

Thank you very much for taking the time to consider my request.

Yours sincerely,

Laura Chen

林教授您好：

我正準備申請好事多公司在芝加哥分店的業務助理一職，希望您能幫我寫一封推薦函。

因為曾在您底下工作過好幾年，所以比起其他人，我相信您更了解我的工作態度以及辦事能力。若您能於信中提及我的優勢和長處，一定會大大增加我錄取的機會。

隨信附上我的履歷給您參考，煩請您聯絡好事多公司的副總格林小姐。若您需要其他資訊，請再讓我知道。

非常謝謝您考慮我的請求。

陳蘿拉 敬上

key words ▲ E-mail 關鍵字一眼就通

1 process 名 過程

2 seek 動 尋找；探求

3 corporation 名 股份有限公司

4 attitude 名 態度

5 strength 名 長處

6 potential 形 潛在的；可能的

複製 / 貼上萬用句 Copy & Paste

I am writing in the hope of your assistance to write a reference letter for me.	我希望您能幫我寫推薦信。
I am writing to ask whether it would be possible for you to provide a reference letter for me.	我來信是為了詢問您，是否方便替我寫一封推薦信。
My contact to you today is regarding a recommendation I am seeking for you to write on my behalf.	我想請您代表我寫推薦信。
I'm applying for a marketing position at ABC Ltd., and am hoping that you would consider writing a letter of recommendation for me.	我打算申請 ABC 股份有限公司行銷主管的職位，希望您能幫我寫推薦信。

I am writing to ask if you would be willing to write a letter of recommendation for me for my application to ABC Company.	我來信是想請問您是否願意幫我寫推薦信到 ABC 公司。
As you are a highly respected teacher of mine, your opinion must be decisive for the company I am applying to.	因為您是我非常尊敬的老師，所以您的意見必定會受到我應徵公司的高度重視。
Therefore, I am hoping that you would agree to serve as one of my references.	所以希望您能夠同意幫我寫推薦信。
Since we worked together for three years, I feel that you are best qualified to offer this information.	因為我們曾一起工作三年，所以我相信您最能擔任我的推薦人。
Please let me know if you will be able to draft a recommendation letter on my behalf.	如果您方便替我寫推薦信的話，請讓我知道。
Please let me know if you will be able to assist with my searching for employment.	在求職這一方面，如果您能幫忙我的話，請讓我知道。
If you would be willing to do so, please briefly describe any qualifications which you feel would make me a good candidate in the position.	如果教授願意幫我寫推薦函的話，麻煩在信上簡短描述我有什麼足以勝任這份工作的特點。
Thank you in advance for your kind attention to these requests. Should you have any questions, please do not hesitate to contact me.	謝謝您重視這個請求，如果您有任何問題的話，歡迎隨時與我聯絡。

3-02 工作推薦信函

推薦信（Recommendation Letter）包括：一、推薦人與被推薦人認識與共事的時間；二、推薦人（老闆、朋友、同事、教授等）與被推薦人在職場或專業上的關係；三、推薦人對於被推薦人的個性、資格及工作能力的評價。

From:	Henry Black
To:	ABC INC.
Subject:	Recommendation Letter

To Whom It May Concern:

It is my pleasure to recommend Mr. Charles Wang for **employment**[1] with your organization. Charles has been working as an **accountant**[2] in our office for three years, and I have been his **supervisor**[3] all these years. Charles' service as an accountant has been **eminently**[4] satisfactory in every way. He is careful and **conscientious**[5] about his work, and is also pleasant toward clients and coworkers.

Charles is planning to find new opportunities to **expand**[6] his **capabilities**[7] in the accounting industry and for **broader**[8] career **horizons**[9]. I highly recommend Charles for employment without reservations. I'm confident that he will be a great asset to any organizations and will be successful in any task given.

If I can be of any further assistance, please feel free to contact me at the phone number below.

Yours sincerely,
Henry Black
Y&I Corp.
Tel：(02) 1234-5678

敬啟者：

我很榮幸向您推薦王查理先生到貴公司任職。查理擔任本公司會計已有三年的時間，我是他的主管。他在各方面都表現極佳。對於工作，查理很謹慎且負責，個性也很好。

查理之所以離開公司，是因為他想拓展自己的能力，並往更廣的領域發展。我誠心地向您推薦這位員工，不管面對任何交辦的事務，我相信他都能妥善處理，成為貴公司的重要人才。

若您需要任何進一步的資訊，請使用下面的電話與我聯繫。

Y&I 企業 亨利・布雷克 謹上
聯絡電話：(02) 1234-5678

key words ▲ E-mail 關鍵字一眼就通

1 **employment** 名 雇用　　2 **accountant** 名 會計師　　3 **supervisor** 名 主管

4 **eminently** 副 突出地　　5 **conscientious** 形 認真的　　6 **expand** 動 發展

7 **capability** 名 能力　　8 **broad** 形 廣泛的　　9 **horizon** 名 範圍

複製 / 貼上萬用句 Copy & Paste

第一段：開門見山的推薦

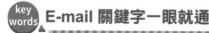

It is a pleasure for me to recommend Ms. Chen Mei-ing for employment with your company.	推薦陳美鶯女士到貴公司工作是我的榮幸。
I am proud to recommend a beloved secretary of mine, Kaya Stone, for your fellowship partner.	非常驕傲能推薦我鍾愛的祕書卡雅・史東，作為您的合夥員工。
I would like to take this opportunity to recommend Rita to your company.	我想藉此機會向您推薦芮塔到貴公司。

I am pleased to comply with Nancy's request, for I think she is well qualified as a secretary.

我很樂意推薦南西,她非常適任貴公司的祕書一職。

第二段:描述表現及能力

She has been working in our company for nearly two years; and she clearly stood out among her coworkers.

她在我們公司工作了將近兩年的時間,一向都比其他同事還要來得出類拔萃。

It has been a great pleasure for me to work with Mr. White, our marketing officer from 2012 to 2017.

我非常榮幸能與懷特先生在二〇一二年到二〇一七年間於公司的行銷部門共事。

He is sincere, positive, and ambitious.

他為人誠懇、積極,且很有企圖心。

He has proven himself to be capable of managing and directing by his work ethic and attitude.

他的工作態度證明他是一個有管理才能的員工。

Jane has been able to establish a good rapport with her colleagues.

珍生性樂觀,與公司的同事互動良好。

She has proven to be a go-getter at work who always plans ahead and gets things done perfectly.

她在工作上幹勁十足,總會做好規劃並付諸實行。

She has successfully demonstrated her leadership by counseling newcomers in her department.

輔導部門新進員工時,她充分展現她的領導才能。

Bill has outstanding organizational skills.

比爾具備卓越的組織能力。

He has the ability to take on any challenge he faces.

不管遇到什麼挑戰,他都有能力解決。

I'm proud of him as his teacher, and I am sure he will continue to excel in your department.	身為他的老師,我感到非常驕傲,我確信他未來會在您的部門發光發熱。
Carl is a highly intelligent young man with high capacity for understanding.	卡爾是一位非常聰明、領悟力高的年輕人。
I highly recommend him for any positions.	我極度推薦他。
His leadership, extraordinary educational background, and many other advantages prove him capable for any tasks.	他的領導力、學歷、以及許多其他的優勢足以讓他勝任各種工作。
I am confident to recommend Kaya as an excellent writer and publicist, judging from her tenure here.	卡雅曾任職於我們機構,我可以非常肯定地向您推薦,她是一位優秀的作家及出版者。
She is confident, and is able to handle any tasks with maturity and enthusiasm.	她相當有自信,能夠以成熟的態度與熱忱處理事情。

第三段:收尾,再次推薦

With his great capabilities and enthusiasm, I believe that he will be a great asset to your organization.	他的能力出眾,並充滿熱忱,我相信他會成為貴公司的重要人才。
I am confident that he will continue to succeed at work.	我相信他未來在工作上會有傑出的表現。
As he will focus on languages development in the future, I truly believe that he would be an asset to your company.	未來他會持續拓展語言能力,我相信他一定能成為貴公司的助力。
I am confident that Mr. Jackson is qualified as a senior manager.	我相信傑克森先生能勝任高級主管一職。
I am confident that Linda will continue to advance in her research.	我相信琳達會持續在她的研究上發光發熱。

Part
1
入門篇

Part
2
求職篇

Part
3
客戶往來篇

Part
4
人際互動篇

- It is for these reasons that I recommend Hannah without reservations.

這正是我毫無保留地推薦漢娜的原因。

- Angela comes with my highest recommendation.

我極力推薦安琪拉。

- I highly recommend Daniel for employment.

我力薦丹尼爾就任。

- I recommend Danny Chou to your company with absolute confidence.

我有絕對的信心推薦周丹尼到貴公司。

- Thank you once again for the opportunity to recommend such a special and impressive young man.

再次感謝您給我機會推薦這位特別又令人印象深刻的年輕人。

提供後續聯絡方式

- If you need more information about Miss Thompson, please do not hesitate to contact me.

如果您需要湯普森小姐進一步的資料，請與我聯繫。

- If you have any questions regarding this recommendation, please do not hesitate to contact me.

如果您對這封推薦信有任何疑問，請與我聯絡。

- If there are inquiries about her performance and characteristics, I would be more than happy to provide further information.

對於她的工作表現與人格特質，若您有疑問的話，本人定當樂意回覆。

- I am looking forward to receiving your favorable reply.

我非常期待您的回覆。

Unit 4 徵才與求職
To Find a Job

4-01 求職博覽會與求職信

各大專院校每年會與企業合作舉辦求職博覽會。對於感興趣的工作，可以利用英文求職信增加機會。如果能寫出一封內容完善、表達清楚合宜的求職信，就能讓雇主了解你的能力，進而留下深刻的印象，幫助你取得面試機會與職位。

From:	Elaine Moss
To:	XYZ Inc. Personnel
Subject:	National Career Fair

Dear Sir:

I've joined the National Career **Fair**[1] on **March**[2] 25 and seen the **post**[3] of Executive Secretary vacancy. I am quite interested in this position.

I have enclosed my personal resume with **pleasure**[4]. As you will find in my resume, I have been working as a **private**[5] secretary to Mr. Chou, sales director at Hannstar Corp. Ltd. for the past five years. Therefore, I am confident that I can **carry out**[6] all **duties**[7] of an executive secretary with a **satisfactory**[8] quality. If I have the honor to work for you, I'll offer my best with **loyalty**[9].

I am available for an interview at your convenience, and can be reached at 02-3088-8855. Thank you for considering my request.

Best regards,

Elaine Moss

敬啟者：

三月二十五日我參加就業博覽會時，看到貴公司有關執行祕書的職缺，感到十分有興趣。

我很樂意附上我的個人履歷給您參考。如您所見，我曾在瀚宇彩晶公司擔任業務處長周先生的私人祕書長達五年，我相信這些經驗足以讓我勝任執行祕書一職。若有這個機會在貴公司服務的話，我必會盡心盡力地完成工作。

不管什麼時候，只要您方便，我都能配合公司時間參加面試。在此附上我的聯繫電話：02-3088-8855，非常感謝您撥空閱讀這封信。

伊蓮・摩斯 敬上

key words ▲ E-mail 關鍵字一眼就通

1 fair 名 博覽會　　　2 March 名 三月　　　3 post 名 職位

4 pleasure 名 愉快　　5 private 形 私人的　　6 carry out 片 執行

7 duty 名 責任　　　　8 satisfactory 形 令人滿意的　9 loyalty 名 忠誠

實力大補帖 Let's learn more!

深入學 ▲ 資訊深度追蹤

📍 常見的求職管道

01 人力銀行與公司網站：人力銀行是非常普遍的求職管道，不過就大公司而言，還有一個很常用的方式，就是直接在公司網站上招募人才，想要求得機會的人不妨時時注意，以免錯失良機。此外，如果想要應徵海外公司的職缺，就要選擇該國熱門的人力銀行刊登履歷，才不會寫了半天的履歷卻無人問津。

02 人力仲介（Headhunter）：人力仲介雖然在國內比較少見，但在國外卻是很普遍的一種求職管道。一般而言，大公司多半有自己配合的人力仲介，仲介會依據最後談成的薪資，向公司收取費用。不過，人力仲介的目標一般為有實力、具兩年以上工作經驗的人才，求職新鮮人多半不在仲介的目標範圍內。

03 內部引薦：如果認識的人脈當中，有在業界服務的對象，可以尋求對方的協助，替自己引薦。若人脈為有力人士，甚至可大幅度地提升面試機會。

　　求職時，一般人往往只注重寫履歷，而忽略了求職信（Application Letter / Cover Letter）。簡單來說，求職信是吸引雇主給予面試機會的重要工具。內容需包括：一、表明自己對職缺的興趣；二、簡要說明自己的資格、工作經驗、能力等；三、表達希望能有面試機會，並留下聯絡方式。

　　若不知道收件者的姓名，可以用「Dear Sir or Madam」來稱呼。但無論收信者是誰，求職信都一定要引人注意，把內容的重心放在自己的相關經驗與能力上，並表示自己的強項能符合公司未來的需求。此外，求職信的最後一段可以告訴雇主你將採取什麼行動，藉此表現積極的態度，以增加面試機會。

From:	Blake Wood
To:	Ms. Gill
Subject:	Apply for Web Designer Position

Dear Ms. Gill:

I am **applying for**[1] the web designer **position**[2] which was **advertised**[3] on your website this week.

This position requires skills in programming and software, and I have experiences in webpage designing, especially in web systems and **operations**[4], thanks to my **previous**[5] working experience. I have enclosed my resume for more detailed information.

Would you consider my request for a personal interview to further discuss my **qualifications**[6]? I can be reached at 1234-5678 or blakewood@mail.net. Please feel free to contact me anytime. I will also **make a call**[7] to see if a meeting can be arranged.

Thank you **in advance**[8] for your **consideration**[9] on my request. I look forward to talking with you in person.

Sincerely yours,

Blake Wood

Part
1
入門篇

Part
2
求職篇

Part
3
客戶往來篇

Part
4
人際互動篇

吉爾女士您好：

我想要應徵本週刊登於貴公司網頁上，有關「網站設計」的職缺。

貴公司要求程式設計和軟體技術的技能，而我剛好在前一家公司學會各種網頁設計的操作技巧。隨信附上個人履歷，提供您更詳細的資料。

可否提供我面試的機會，以便進一步討論我的資格與機會？如果需要與我聯繫，請隨時來電：1234-5678，或來信：blakewood@mail.net.，我也會去電與您確認面試機會。

謝謝您的考量，期盼與您會面。

布雷克・伍德 敬上

 E-mail 關鍵字一眼就通

1 apply for 片 申請　　2 position 名 職位　　3 advertise 動 刊登廣告

4 operation 名 操作　　5 previous 形 以前的　　6 qualification 名 資格

7 make a call 片 打電話　　8 in advance 片 預先　　9 consideration 名 考慮

實力大補帖 Let's learn more!

 文法 / 句型解構

📍 求職信實用句

1 **I'm writing in response to your recently advertised position for a...** 我來信是為了您所刊登…的職缺廣告。（空白處填寫「職位名稱」）

2 **I came across your profile on LinkedIn and saw that you are a recruitment specialist in...** 我在 LinkedIn 看到您的個人檔案，注意到您正在替…應徵新人。（空白處填寫「部門單位」）

3 **Enclosed is my resume that more fully describes my background and work experience.** 隨信附上我的履歷，關於我的背景與工作經驗，履歷上面會有更詳細的說明。

4 **My key skills are ... and my career highlights include:** 我最主要的優勢是…，針對這個工作，我還有其他的相關能力，包含：

撰寫求職信（Cover Letter）時，一定要注意「求職信需量身訂做」，千萬不要亂槍打鳥，應針對職缺，強調與其相關的教育背景、工作經驗、以及相關能力。

在描述優勢的同時，要注意內容須簡短有力，建議篇幅不要超過一頁 A4 的長度。

From:	Grace Huang
To:	Sunny-Smile
Subject:	Summer Job Opportunities

Dear Sir:

I am writing to **enquire**[1] if you have any **opportunities**[2] for children's **entertainers**[3] and nannies at your **institution**[4] this summer.

I am twenty years old and am currently pursuing a degree in Tourism at the College of Travel and Tourism in Singapore. I had also been working as a part-time assistant at a local kindergarten. I am fully experienced in **looking after**[5] babies and young children up to the age of six, and have **expertise**[6] in babysitting, preparing children's meals, and organizing activities. I enjoy looking after children and I like working as part of a team.

My **mother tongue**[7] is Chinese and I have a good **command**[8] of spoken English. I also speak a little Spanish.

I would be grateful if you could send me details of any available positions as well as an application **form**[9]. Thank you in advance for considering my request.

Sincerely yours,

Grace Huang

Part
1
入門篇

Part
2
求職篇

Part
3
客戶往來篇

Part
4
人際互動篇

先生您好：

我寫信來詢問貴校今年暑假是否有提供兒童指導員或褓姆的工作？

我今年二十歲，目前在新加坡的旅遊及觀光學院就讀旅遊學程。我曾在當地的幼稚園兼任助理，因此對照顧幼兒和六歲以下的孩童很有經驗。我能夠勝任褓姆的工作，也很會準備餐點和規劃活動。我喜愛照顧小孩，也很喜歡在一個團體裡面工作。

我的母語是中文，英文口說也很流利，除此之外，我還會說一點西班牙文。

若您可以寄其他詳細的工作職缺或申請表給我，將感激不盡。

葛蕾絲‧黃 敬上

key words ▲ E-mail 關鍵字一眼就通

1 enquire 動 詢問　　　2 opportunity 名 機會　　　3 entertainer 名 表演者

4 institution 名 機構　　4 look after 片 照顧　　　6 expertise 名 專長

7 mother tongue 片 母語　8 command 名 運用能力　　9 form 名 表格

 實力大補帖 Let's learn more!

深入學 ▲ 資訊深度追蹤

● 非「避」不可的 NG 求職信

　　大部分的應徵者都會在履歷上面下足功夫，所以決定勝負的往往就是求職信。但是，要讓求職信成為關鍵，就必須避開撰寫時的雷區，才不會白白讓求職信成為扣分元素，以下就列出求職者較常犯的幾項錯誤。

01 搞不清楚「求職信」與「自傳」的差異：許多外商公司往往更重視求職信，下筆前一定要弄清楚公司要求的是哪一份文件，千萬別把求職信當自傳寫。

02 粗心大意的錯誤：要是把應徵職位（甚至公司名稱）寫錯，給人的印象可是會大打折扣的。

03 與履歷（Resume）的重複性太高：重複性過高的內容，只會讓雇主認為你的專業不夠廣泛。

4-04 人脈求職信

人脈求職信（Networking Letter）和一般求職信的不同在於「開頭的緣起」，也就是「如何得知公司的徵才訊息」。一般求職者多半提及的會是報紙、網站等媒介，但若你有人脈，可於求職信中提及此人，讓雇主對你有更深的印象。

From:	Emma Wilson
To:	Mr. Hamilton
Subject:	Career Decision

Dear Mr. Hamilton:

Dr. Steven Johnson, professor of business at City University, suggested that I make a contact with you. He thought that you, as an excellent alumnus, would be able to assist me with career decisions.

As a business major, I'm interested in both marketing and economics, and am still exploring my future career. I'd like to attend a campus interview next semester, and would like to seek for your help on career advice.

May I call you next week to see if we can arrange a meeting at your convenience? Thank you for considering my request.

Sincerely,

Emma Wilson

漢米爾先生您好：

城市大學的商學教授史帝芬‧強森先生建議我與您聯繫。他表示以您優秀校友的身分，可以幫助我規劃我的職涯。

身為一名商學院學生，我對行銷和經濟學都很有興趣，正在思考未來職涯該往哪個方向發展。下學期我準備參加校園面談會，為此我想要向您尋求對職業選擇的建議。

若您下週有空的話，是否能去電詢問您方便的會面時間呢？感謝您的幫助。

艾瑪‧威爾森 敬上

Part
1
入門篇

Part
2
求職篇

Part
3
客戶往來篇

Part
4
人際互動篇

4-05 針對公司的求職信

針對公司的求職信（Prospecting Letter）就是投遞到某公司，詢問對方是否有職缺的信件（並不確定對方有職缺）。和一般求職信不同的是，這類信件的重點在「自己很能適應該公司的環境」，而非針對某個職位而寫。

From:	Elle Zheng
To:	Mr. Anderson
Subject:	Retail Management Positions

Dear Mr. Network:

I read about your company in US Job Career **magazine**[1], and would like to inquire about employment opportunities in your management training **program**[2], especially in **retail**[3] management, which I'm interested in.

I will receive my B.S. **degree**[4] this June in Economics. I became interested in business when I was in high school, and took a variety of jobs related to retail during college. The professional working environment in your organization is just the kind that I'm seeking.

Please find my resume in the **attachment**[5]. I believe my educational background and working experiences meet your **requirement**[6], and I have strong **interpersonal**[7] **skills**[8] and **ambition**[9] to build a successful career in retail management.

I know how busy you must be during this time of year, but I would appreciate a few minutes of your time. May I call you during this week to discuss any employment possibilities? In the meantime, you can also contact me at 1234-5678 or ellezheng@mail.net.

Thank you very much for considering my request. I look forward to talking with you.

Sincerely,

Elle Zheng

安德森先生您好：

我看到貴公司刊登在《US求職雜誌》中的內容，因此想詢問有關管理訓練部門的工作機會。尤其是零售管理類的工作，我對於這個領域特別有興趣。

今年六月我將於經濟系畢業。我對商業的興趣源自於高中，大學時期則從事各式各樣的零售工作，而貴公司正好提供我所追求的專業環境。

隨信附上我的履歷供您參考。我的學歷和經驗都符合貴公司要求，我很善於處理人際關係，這能讓我在零售管理的領域上有很好的發展。

我想您此時必定十分忙碌，但若能給我面試的機會，將不勝感激。如果方便的話，本週我將致電貴公司，與您討論工作機會。您同時也可以透過電話和電子郵件聯繫我，聯繫方式如下：

電話：1234-5678；電子郵件：ellezheng@mail.net

謝謝您的考量，衷心期待與您見面。

鄭艾兒 敬上

key words ▲ E-mail 關鍵字一眼就通

1 magazine 名 雜誌　　2 program 名 計畫　　3 retail 名 零售

4 degree 名 文憑　　5 attachment 名 附件　　6 requirement 名 要求

7 interpersonal 形 人際間的　　8 skill 名 能力　　9 ambition 名 抱負

複製 / 貼上萬用句 Copy & Paste

開頭：看到求職公告

I am writing in response to your advertisement in the Career Journal for a Product Manager.	我來信回應貴公司在《職涯月刊》上所刊登的產品經理徵才廣告。
Your advertisement for an Engineer posted in the Career Journal is of great interest to me.	我對貴公司在《職涯月刊》上所刊登徵求工程師的廣告十分感興趣。

I am a recent university graduate of Warwick Business College and am interested in starting a career in your organization.

我剛從華威商學院畢業，對於在貴機構展開個人職涯深感興趣。

I've seen your advertisement in China Times of July 12 about a position of clerk in a warehouse in Hsinchu. I'd like to offer myself for the post.

七月十二日我在《中國時報》看到您刊登的廣告，想要詢問並申請新竹批發店的店員職缺。

Your advertisement for a production management at the Job Fair is interesting to me, because your requirements closely parallel my working experiences.

我在求職博覽會上看到貴公司的應徵說明，對生產管理的職位深感興趣，而且我的工作經驗和您要求的條件十分相符。

I am writing in response to your advertisement at the Job Fair on March 11. I am quite interested in the position you offered in your department.

我寫信回覆有關於三月十一日在求職博覽會中的廣告，我對您部門所提供的職缺深感興趣。

The sales position which you described in your advertisement at the Career Fair of March 11 is the one that I think I can show you some excellent qualifications.

我在三月十一日的求職博覽會上看到貴公司的廣告，我認為自己的條件十分符合貴公司所要求的業務資格。

TinTin Yang of TTV Enterprise suggested me to contact you regarding a vacancy in the marketing department.

TTV 企業的楊婷婷建議我向您詢問有關行銷部的職缺。

After speaking with John Harris, an officer of your Vice President Department, I was excited to learn that there's a match between your needs and my qualifications.

與貴公司副總經理室的專員約翰·哈里斯先生談完之後，我很高興得知自身的能力符合你們的需求。

開頭：針對特定公司自薦

I am looking for a position as a sales assistant.

本人正在找一份業務助理的工作。

Perhaps there is a vacancy in your organization for an experienced and conscientious secretary?

請問貴公司是否需要一名有經驗、負責任的祕書？

Would you need an experienced receptionist in your hotel next summer?

貴飯店明年暑期是否需要一名有經驗的櫃台人員？

I'd like to start from the bottom, so any entry-level job is great for me.

我想先從基層做起，所以願意接受任何初階的工作。

Because I am desirous for some accounting experience during summer time, I am writing to inquire whether you will need a young talent like me.

我很渴望於暑假期間得到會計的工作經驗，所以特別寫信來詢問貴公司是否需要我這樣的年輕人才？

自我介紹：年齡和經驗

I am twenty-six years old, female, and have one year sales experience in ABC Company.

我今年二十六歲，女性。曾於 ABC 公司擔任一年業務。

I am twenty-five years old, and have been employed for two years by the England Furniture Co., as an assistant in the marketing department.

我今年二十五歲，曾於大英傢俱公司服務兩年，於行銷部擔任助理。

I am 25 this year, and have been working in my present position for five years.

我今年二十五歲，在目前的工作崗位已有五年的經驗。

I can offer you my references for my professional experiences in the past six years.

我可以提供過去六年間的推薦信，以供您參考。

I am twenty-two years old. I graduated from Soochow University with a major in English Literature.

我今年二十二歲，畢業於東吳大學英國文學系。

Since my graduation in 2015, I have been a secretary in a trade company.	自從二〇一五年畢業後，我一直在貿易公司擔任祕書。
I have been in the business field for five years, and worked as a secretary in the personnel department.	本人過去五年來在商業界擔任人事部祕書。
I have two years' experience as a salesman.	本人曾擔任業務長達兩年的時間。
For these three years, I have been working at the Konomi Food Company as an accountant.	本人在相撲手食品公司當了三年的會計，現仍在職。
Over the years, I have become quite experienced as a departmental coordinator.	經過這幾年的工作，我對跨部門的溝通協調很有經驗。
I am good at English business letters and reports.	我擅長寫英文書信和報告。
I possess a thorough marketing knowledge and am familiar with all product lines and their marketing strategies.	我對於行銷方面的知識瞭若指掌，且通曉各項商品及行銷策略。

敘述個人能力

My former employer has praised me for my ability to meet strict deadlines.	我之前的雇主對我準時完成工作的能力讚賞有加。
My responsibilities included serving international clients and supervising a team of ten.	我的職責包括服務國際客戶以及督導十位同仁。
I managed a regional marketing campaign which increased customer interest in our products and services by 20%.	我掌管一項區域性的行銷活動，使顧客對我們產品及服務的興趣提高了百分之二十。

🖱	During the years at HannStar Corp., I saved the company over USD 300,000 through various cost-cutting strategies.	在彩晶公司服務的期間，我透過各種降低成本的策略為公司省下三十幾萬美金。
🖱	My experiences at HannStar Corp. make me confident to meet the aggressive sales objectives you have set.	彩晶公司的工作經驗使我確信自己能夠達到貴公司定出的積極銷售目標。
🖱	As the enclosed resume indicates, I have more than ten years' sales experiences in all kinds of field.	從附上的履歷表可知，我擁有十年各種業務方面的工作經驗。
🖱	For the past five years, I've supervised a group of at least ten employees.	過去五年來，我管理一個包含十幾位成員的團隊。

其他技能與特色

🖱	During my tenure as a counselor, I developed communication and interpersonal skills.	我在擔任顧問期間，培養出溝通與人際相處的技巧。
🖱	My language skills in English and French will prove valuable in your goals to expand into the European countries.	您目標進軍歐洲市場，而我的英文與法文能力將能替貴公司帶來無法估量的效益。
🖱	I spent two years in Germany as an exchange student; therefore, I speak and write German very well.	我去德國當了兩年的交換學生，德文說寫流利。
🖱	I am very efficient under tight deadlines. In fact, I perform better when under pressure.	當時間很緊迫的時候，我的工作效率也很高。事實上，我在壓力下的表現更為出色。

請求面試與感謝

I'd appreciate the chance to talk to you in person and to better know whether I have the chance to get the job you offered.

若能有機會與您會面，了解是否有機會取得公司所提供的職缺，將不勝感激。

I would appreciate the opportunity to meet you in person and to discuss how I can contribute to your company.

若有機會和您親自見面，討論我如何在貴公司貢獻所長，我將非常感激。

I would be delighted to meet you and to discuss how I can fulfill your marketing needs.

我很樂意與您見面，討論我能如何滿足貴公司的行銷需求。

I would be pleased to come by your company and to go over my resume and teaching methods.

若能前往貴公司，就我的履歷和教學方法進行討論的話，我將非常高興。

This job sounds interesting because it is the kind of work I have wanted to do for years.

我對這份工作很感興趣，因為這是我多年來一直想要找的工作。

I hope you will give my application due consideration.

希望您可以慎重考慮我的申請。

If I can provide you with any other necessary information about myself, please don't hesitate to let me know.

若您需要我提供更多個人資料，請隨時通知我。

I will be happy to tell more about my experiences in an interview.

我很樂意在面談中詳述我的工作經驗。

I look forward to discussing my potential to contribute to your clients' needs.

十分企盼能與您討論我未來能如何滿足貴公司客戶的需求。

Thank you for taking the time to consider my application, and I look forward to hearing from you.

感謝您抽空閱讀我的資料，期盼能收到您的回覆。

邀約面試與結果通知
To Attend an Interview

5-01 通知面試時間

面試時間通知信（Interview Notice）通常會以「您的資料已由本公司審核通過，特此通知您面試相關資訊」為首發展；具體內容則包括面試時間、地點及聯絡人的資料。

From:	Duke White
To:	cindy.c@mail.net
Subject:	MIT Interview Notice

Dear MIT Applicant:

I am pleased to inform you that your application to our company is under consideration. We would like to interview you in person and discuss more details in the following location:

26th floor, No. 1, Songzhi Road, Xinyi District, Taipei 110, Taiwan, R.O.C.

Please let me know when you would be available for an interview as soon as possible. The interview will last for 30 minutes. The final result will be announced on our company website by 5:00 p.m., Monday, April 8, 2018.

Please call at 02-2288-5434 between 10:00 a.m. and 4:00 p.m. to schedule your 30-minute interview.

We look forward to seeing you.

Best wishes,

Duke White

Manager, Human Resources Department
MIT
ext. 2601

Part
1
入門篇

Part
2
求職篇

Part
3
客戶往來篇

Part
4
人際互動篇

親愛的 MIT 應徵者：

我們審核了您應徵本公司的資料，想要通知您前來面談，地點如下：

110 台北市信義區松智路 1 號 26 樓

請盡快告知方便面試的時間。面試時間為三十分鐘。最後的錄取結果將會於二〇一八年四月八日（星期一）前，發布在本公司的網站上。

請於上午十點至下午四點之間來電，確認您方便面試的時間。連絡電話為：02-2288-5434。

我們期待與您面談。

杜克・懷特 謹上

人資部經理
MIT 企業
分機號碼 2601

複製 / 貼上萬用句 Copy & Paste

We have reviewed your resume, and would like to discuss your qualifications in person.	我們已檢視您的履歷，希望與您當面討論您的資歷。
Your qualifications appear to be a good fit for this position. Please contact me to arrange for an interview.	從您的資歷看來，您似乎能勝任此職位。請與我聯繫以安排面試。
We are evaluating applications and will invite candidates to interview on April 26, 2018.	我們正在審核應徵者的資料，且將邀請錄取候選人參加二〇一八年四月二十六日的面試。
Our interview preparation guide is available on the website.	面試的準備提醒已公告在本公司的網頁上。
It is notified for the information of the candidates that our Staff Selection Commission will conduct interviews as scheduled below:	這是通知應徵者的訊息，我們的員工遴選委員會將舉行面談，面試時間表如下：

It is notified for the information of all interviewers that:	這是通知所有面試者的訊息：
The interview of the candidates who have applied for the post shall be conducted on July 4, 2018 at Camp Office of the Company from 8:00 a.m. onwards.	面試將於二〇一八年七月四日上午八點開始，請應徵者於當日前往本公司的 Camp 辦公室。
Please dress in business formal attire.	請穿著正式服裝面試。
As a reminder, please provide an official copy of your resume to your interviewer.	提醒各位，請準備一份履歷表給面試官。
It is also made clear that our Staff Selection Commission will not be responsible for any delay.	也請注意，我們的員工遴選委員會對應徵者自身的延誤不負任何責任。
When you arrive for your interview, please check in with the receptionist, and you will be told where to find the interviewer Ms. Smith.	當你抵達面試會場後，請向櫃檯人員報到，他會告訴你與史密斯女士的面試地點。
Before leaving for your interview, please check your e-mail for urgent messages from the MIT Company, such as delays or cancellation due to weather or emergencies.	前往面試會場前，請確認電子信箱是否有來自本公司的緊急通告，例如：因天候不佳或其他意外而必須延期或取消面試。

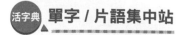 實力大補帖 Let's learn more!

活字典 單字 / 片語集中站

📍 強調自身的優勢

problem solver 問題的解決者	communicator 溝通者
negotiator 交涉者	specialist 專家

initiative 進取心	motivate 激發；刺激
accomplish 實現	analyze 分析
raise the standards 提高標準	see the big picture 著眼大局

 ## 資訊深度追蹤

📍 面試注意事項

好不容易爭取到面試的機會,當然要做一些準備,好替自己在面試時爭取更多印象分,以下就提供幾個加分的方向。

一、面試前

01 做好功課:預先了解對方公司的產品、服務、管理、文化等。

02 做足練習:在朋友、家人或是鏡子面前做面試演練。

03 提早十分鐘抵達面試現場,並攜帶備份履歷表、推薦信及個人作品。

二、面試過程

01 因為第一印象很重要,所以須穿著合適服裝,並以自信的態度與對方握手。

02 與面試官對談時,須避開「一問一答」的模式,盡量營造自在溝通的氛圍。

03 過程中須展現出熱忱,讓面試官了解你對這份工作的熱情。

三、面試結束後

01 建議索取面試官的個人名片,並於面試後立即發送感謝郵件給對方。感謝信的內容,除了謝謝面試官抽空與你面試之外,最重要的是再摘要式地總結自己的優點與能力,強調自己之所以適合這份工作的原因,期能再一次獲得注意。

02 記錄面試人員的姓名和職位,並記錄此份工作的資料,最後仔細評估自己是否能夠勝任這份工作。

5-02 錄取通知信

錄取通知信（Employment Notification）為通知應徵者通過面試之信函，編寫時，除了恭喜應徵者外，內容還應該：一、通知報到的時間及地點；二、提醒對方需準備的物件；三、薪資和福利的確認文件。

From:	Terry Jackson
To:	Mr. Tseng
Subject:	Offer Letter from Y&I Company

Dear Mr. Tseng Yi Lin:

Congratulations! It is my pleasure to offer you **employment**[1] with Y&I **Textile**[2] Company. Your title will be Sales Manager, and your employment will begin on November 20, 2018.

The whole team looks forward to working with you. We believe that you would be a great fit for this position with 8 and half years of experience, which would **contribute**[3] in setting new **strategies**[4] and **targets**[5] in our company.

You will receive a base salary of USD 80,000 per year, and your annually benefits are included in the **enclosed**[6] file.

For any further inquiries, please contact me at below number.

Best wishes,

Terry Jackson

Manager, Human Resources Department
Y&I Textile Company
TEL: 3302-4567 ext.306

曾奕麟先生您好：

我們正式恭喜您，取得 Y&I 紡織公司的業務經理一職，您的到職日為二〇一八年十一月二十日。

我們全體同仁都非常期待您的就任。我們相信您是這職位的最佳人選，且您長達八年半的工作經驗肯定會很有幫助，相信您將能替公司設立新的策略方向，朝新目標邁進。

最後，本公司將提供您八萬元美金的年薪。關於您的福利，請參考附檔文件。

若您有任何疑問，可以來電詢問我。

泰瑞・傑克森 謹上

人力資源部經理
Y&I 紡織公司
連絡電話：3302-4567　分機號碼 306

 E-mail 關鍵字一眼就通

1 employment 名 雇用　　2 textile 名 紡織品　　3 contribute 動 提供

4 strategy 名 策略　　　5 target 名 目標　　　6 enclosed 形 封入的

 複製 / 貼上萬用句 Copy & Paste

通知錄取

It is with great pleasure that we offer you the position of Product Manager.	很榮幸我們將提供您產品經理一職。
We look forward to having you as Executive Secretary on September 1st.	我們期待您自九月一日開始接任公司的執行祕書。
I am very pleased to offer you the position of Sales Assistant.	我很高興提供您業務助理的職位。

🖱	This is a formal job offer for the position of Deputy Director, starting from September 1st for a period of one year.	正式通知您獲聘副處長一職，自九月一日起為期一年。
🖱	As per our phone conversation, the following is a formal job offer of the position of Sales Manager.	如我們在電話裡談的，正式通知您獲聘業務經理一職，以下為詳細內容。
🖱	On behalf of the department, I welcome you as a new employee of us.	謹代表我們部門，在此通知您成為本公司的一份子。
🖱	I am pleased to tell you that you have passed all our qualifications and requirements for the Executive Secretary position.	本公司很高興在此通知，您已通過我們的資格要求，錄取執行祕書一職。
🖱	You will be working with seven other officers in the Vice President Office.	您將在副總經理辦公室與七名專員一起工作。
🖱	As president of the Young Leaders Society, I am proud to inform you that you have been accepted as a new member of our organization.	身為青年領袖協會的總裁，我很高興通知您成為我們組織的一員。
🖱	We have carefully screened all applicants and we deemed your application to be worthy for acceptance.	我們已謹慎審核所有應徵者的資料，認為您是值得我們錄取的一位。
🖱	Your application immediately caught our attention as you have exceptional grades and are also the president of the student council.	您所附上的申請資料，不管是優秀的成績還是曾任學生會主席的經歷，都很吸引我們的注意。
🖱	Your personal characteristics match our needs, and we believe you will bring more success to our organization.	您擁有本公司所需要的特質，我們相信您將會為本機構帶來更多的效益。
🖱	Please attend the formal induction on January 1st, 2018 at the organization office. Welcome to our company!	請於二〇一八年一月一日參加新員工的就職儀式。歡迎加入本公司！

We are pleased to confirm you're being employed by our firm as Sales Manager.

我們很開心通知您受雇為本公司的業務經理。

You will report directly to Vice President, commencing with your start of employment on March 19, 2018.

請直接向副總經理報到，任職日期從二〇一八年三月十九日開始。

薪資與福利

You will receive a base salary of NT$28,000 per month.

您的月薪為二萬八千元。

During the first year of employment, employees will accrue a prorated amount of the standard allowance, so you will be entitled to a seven-day vacation for this year.

第一年的休假日將按比例分配，所以這一年會分配到七天假期。

As an employee, you will be eligible for salary increases based upon your performance.

員工可依照工作表現取得加薪。

You are eligible for paid vacation and participation in the company retirement plan.

你擁有不扣薪的年假，並可參與公司的退休計畫。

信件結語

Congratulations! We look forward to having you in our company.

恭喜！期待您加入我們公司。

We look forward to your joining of our company.

我們期待您加入公司。

5-03 不錄取通知信

不錄取通知信（Rejection Letter）的主要目的為通知應徵者未通過面試，因此通常會以「感謝您前來參加面試」為開頭，接著委婉地告知不錄取之原因。信末可給予鼓勵及加上「未來若有機會，仍有合作空間」之訊息。

From:	Jack Brown
To:	Mr. Lai
Subject:	Rejection Letter - Position Filled

Dear Sheng-Yung Lai:

Thank you for applying for the position in PCP Company. We appreciate your time and patience **throughout**[1] the interview process. However, after careful **consideration**[2], we had made a difficult decision to not give you an **offer**[3] at this time. **Nonetheless**[4], we will have your resume in our **database**[5] for reference. If there is any position suitable for you in the future, we will definitely contact you.

We do thank you for your interest in PCP Company, we wish you good luck in your future **endeavors**[6].

Sincerely,

Jack Brown

Manager, Human Resources Department
PCP Company
ext. 501

Part
1
入門篇

Part
2
求職篇

Part
3
客戶往來篇

Part
4
人際互動篇

賴申勇先生您好：

感謝您應徵 PCP 公司的職缺，也很謝謝您前來參與面試，以及期間的耐心。但是，在經過慎重的考慮之後，我們很遺憾通知您，並未錄取該職缺。不過，我們將保留您的資料，若日後有適合的職缺，我們一定會與您聯繫。

謝謝您對本公司的愛護，我們祝福您求職順利。

傑克・布朗 謹上

人資部門經理
PCP 公司
分機號碼 501

 E-mail 關鍵字一眼就通

1 **throughout** 介 從頭到尾   2 **consideration** 名 考慮   3 **offer** 名 提供

4 **nonetheless** 副 但是   5 **database** 名 資料庫   6 **endeavor** 名 努力

複製 / 貼上萬用句 Copy & Paste

信件開頭

It was a pleasure meeting with you to discuss your background and interest in the Sales Manager position in our organization.

很高興能與您面談，討論您的相關背景以及對本公司業務經理一職的興趣。

We appreciate your time, attentiveness and patience throughout the interview process.

我們感謝您在面試過程所投入的時間、專注與耐心。

告知對方未被錄取

We did have several highly qualified candidates for the position, and it has been a difficult decision.

有好幾位合適的人選，令我們難以抉擇。

But we have chosen to pursue another candidate for this position, who we feel is best qualified.	但我們還是選擇了另一位更合適的人選。
We feel that you are better qualified for another position.	我們認為您更適合其他職位。
We would like to forward your application to the other departments for evaluation as well.	我們想要將您的申請資料送到其他部門審核。
Please let us know if you do not agree with this procedure.	若您不同意，請告知我們。
After reviewing your resume, we feel your qualifications would be more suitable for a junior position.	審閱過您的履歷之後，我們感覺初階的職位更適合您。

信件結語

We will contact you if any suitable positions become available.	若有合適的職缺，我們會再與您聯繫。
At this time, we are unable to grant your request for an interview.	我們目前無法滿足您面試的要求。
The position you applied for has already been filled. We will keep your resume on file for six months.	您所應徵的職缺已找到人選，我們會將您的履歷保留六個月。
We do thank you for your interest in our company and we wish you good luck in your future endeavors.	感謝您對本公司的愛護，祝您未來一切順利。

Unit 6 感謝函、接受與拒絕工作信函
To Accept or Reject a Job Offer

6-01 感謝提供面試

於面試結束後發送感謝信，能讓對方留下好印象，進而增加錄取機會。感謝信的內容應包括：一、感謝對方提供面試機會；二、你對這份工作的熱忱；三、強調你將為公司帶來的利益；四、提供對方要求的資料。

From:	Anna Yang
To:	Mr. Thompson
Subject:	Thank-You Letter

Dear Mr. Thompson:

Thank you very much for interviewing me yesterday for the retail management position. I enjoyed meeting you and learning more about the position and Philips in general.

My enthusiasm for the position and my interest in working for Philips were strengthened after the interview. I am confident that my educational background and experiences **match**[1] **perfectly**[2] for this position, and I am sure that I will make a **significant**[3] contribution to the organization.

I'd like to once again **express**[4] my strong interest in the position and in working with you and your staff. This is just the kind of job I'm seeking. Please contact me at 2345-6789 or annayang@mail.net if I can provide you with any **further**[5] information or if you have any other questions.

Again, thank you for your time and for considering me for this **exciting**[6] opportunity.

Sincerely,

Anna Yang

湯普森先生您好：

感謝您昨天提供我零售管理一職的面談機會。我很高興能與您見面，並進一步地認識了貴公司的運作和工作內容。

這次面試加深了我對該職位的熱忱，也更想為飛利浦公司效力。我想我的學經歷十分符合您所要求的條件，也有信心未來能替公司帶來顯著的效益。

我想再次重申我對這份工作以及與你們共事的高度意願，這份工作正是我想要做的。若有其他需要，請隨時撥打 2345-6789 或來函至 annayang@mail.net 與我聯絡。

再次謝謝您的面談與考慮。

楊安娜 敬上

 E-mail 關鍵字一眼就通

1 match 動 和…相稱

2 perfectly 副 完美地

3 significant 形 顯著的

4 express 動 表達

5 further 形 進一步的

6 exciting 形 令人興奮的

複製 / 貼上萬用句 Copy & Paste

It was a pleasure meeting with you and your staff at the company.	能在公司與您和您的員工會面是我的榮幸。
I enjoyed our meeting this morning.	我們今天早上的面談十分愉快。
I was very impressed with the working environment.	貴公司的環境讓我印象深刻。
It was a pleasure discussing the opportunity of Product Manager at ABC Corp.	能與您討論在 ABC 公司擔任產品經理的機會讓我感到很榮幸。
Our talk gave me a clear picture about what kind of sales manager you are looking for.	透過這次的談話，我對貴公司要找的業務經理有了清楚的概念。

I became even more interested in the position since I had the chance to see your operations in person.

在見識到貴公司的運作方式後，我對這個工作更感興趣了。

It is challenging to solve your marketing problems, but I am willing to take the challenge.

解決貴公司的行銷問題很具挑戰性，但我非常樂意接受這個挑戰。

I am particularly interested in your plans for expanding markets to the European countries.

我對您的歐洲市場開發計劃特別感興趣。

As we discussed this morning, my familiarity with the European culture will allow me to better oversee the marketing campaign in European countries.

就如我們早上所討論的，我熟悉歐洲文化，這點讓我更有能力管理在歐洲國家的行銷活動。

I am enclosing my job description which you requested, and have asked my references to send letters of recommendation directly to you.

隨信附上您要求的工作說明，也已經請推薦人直接寄信給您。

I am looking forward to a second interview, which we can further discuss how I can utilize my skills to better the company.

我期待第二次面談能進一步討論如何運用我的專長，來幫助公司成長。

I hope that my qualifications meet your criteria for a Sales Manager at your company.

希望我的條件符合貴公司對業務經理一職的要求。

I appreciate your valuable time and consideration for reviewing my profile and interviewing me for this challenging position.

感謝您撥出寶貴的時間，考慮我的資格，讓我有機會參與面試，爭取這份具挑戰性的工作。

I am very excited to join your organization, and will prove valuable to you.

對於到貴公司服務，我感到躍躍欲試，我會向您證明自己的價值。

　　如果決定接受這份工作，接受工作信函（Acceptance Letter）的內容就必須提及形式上的契約協定，內容包括：一、表示接受工作的聲明；二、簡短說明你接受主管的決定與公司目標；三、你將開始工作的日期。

From:	Owen Well
To:	Ms. Hill
Subject:	Acceptance Letter by Owen Well

Dear Ms. Hill:

I am writing to confirm my **acceptance**[1] of your employment offer of July 4th. I am absolutely **delighted**[2] to join this company; the work is exactly what I love and have been preparing myself to. I am confident that I will make a significant contribution to the corporation, and I am **grateful**[3] for the opportunity you have given me.

As we **discussed**[4], I will start working at 8:00 a.m. on July 15th, and I will complete all employment and **insurance**[5] forms for the new employee **orientation**[6] on that day.

I look forward to working with you and your great team. I appreciate your confidence in me and once again, I am very happy to be part of the team.

Sincerely,

Owen Well

希爾小姐您好：

我寫信來確認接受您於七月四日提供的工作機會，能成為貴公司的一份子讓我備感喜悅。這份工作是我一直所期待的，也為此準備了許多，我有信心可以投身奉獻於公司，同時也很感謝您給我這個機會。

就如我們先前討論過的，我會在七月十五日上午八點到公司參加新進員工訓練，並填寫

員工資料表和保險資料。

非常期待與您和團隊一起工作。感謝您對我的信任，也很高興成為公司的一份子。

歐文・威爾 敬上

 ## E-mail 關鍵字一眼就通

1 acceptance 名 接受　　2 delighted 形 高興的　　3 grateful 形 感激的

4 discuss 動 討論　　5 insurance 名 保險　　6 orientation 名 熟悉

複製 / 貼上萬用句 Copy & Paste

I am pleased to accept your offer of an editor position at Wonderful Books.	我很高興接受萬德福出版社提供的編輯一職。
This letter is regarding the acceptance of your offer received via e-mail dated April 13, 2018.	這封信是要接下您於二〇一八年四月十三日來信提供給我的職位。
Firstly, I would like to thank you for this opportunity.	首先，我要謝謝您願意給我這個機會。
I am extremely happy to take up the position of Assistant Manager.	我十分高興可以得到助理經理一職。
I would like to go ahead with this employment acceptance letter and I am also satisfied regarding the contract period and salary package.	我願意接受這份工作，也十分滿意貴公司所提供的聘期和薪資。
During these five years of my tenure, I will do my best to implement new plans and schemes which can really enhance our business scope.	在這五年的任職期間，我會盡最大的努力來完成新計畫，以擴大公司的獲利範圍。
This is the time for me to prove my capabilities, and I won't let you down.	這是讓我證明自己能力的時候，我不會讓您們失望。

I am very pleased to accept your offer of working together in the field of finance.	我很樂於接受您所提供，從事金融業的工作機會。
I am eager to work with you as soon as possible.	我渴望能盡快與您們共事。
You can be sure I will make good use of my experiences to complete my work.	我將善用我的經驗來完成工作，關於這點您可以放心。
I will be able to start working on Monday, September 1st, as we discussed on the phone.	如同我們在電話裡討論的那樣，我將於九月一日（星期一）開始工作。
I will report to the personnel office at 8:30 a.m., and arrive at office at 9 a.m.	我將於八點半至人事室報到，並於九點鐘抵達辦公室。
Thank you so much for providing me this golden offer.	謝謝您提供我這份絕佳的職缺。
Thank you for this opportunity. I look forward to working with you and the staff soon.	謝謝您給的機會，我期待與您和其他同仁一起工作。

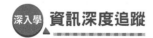
實力大補帖 Let's learn more!

深入學 **資訊深度追蹤**

📍 接受工作信函的內容

接受工作信函用來確認你對這份工作的認同，包括薪資、到職日、醫療檢查等。通常在接到公司的錄取通知後，就能開始著手寫接受工作信函了。一般而言，這種信件可分為三段：

01 第一段：接受，並確認你進入公司的決定。

02 第二段：確認工作的相關內容。

03 第三段：表達感謝之意。

6-03 拒絕工作信函 I

要婉拒公司所提供的工作，就必須寫拒絕工作信函（Withdrawal Letter）。一封能維持好關係的拒絕工作信函很重要，因為日後還是可能會再向該公司應徵。先表示感謝對方提供這個工作機會，再提出拒絕的原因，切記要委婉地表達。

From:	Tracy Scott
To:	Mr. Brown
Subject:	Tracy Scott - Withdrawal Letter

Dear Mr. Brown:

I am writing to inform you that I am **withdrawing**[1] my **application**[2] for the retail management position. As I **indicated**[3] in my interview with you, I have been considering several employment possibilities. This week I was offered an **executive**[4] management position in Chicago and, after careful consideration, I decided to accept it. The position is a good match for me at this point in my **career**[5].

I would like to thank you for interviewing and considering me for this position. I enjoyed meeting you and learning more about the company and the key **projects**[6]. This is a great company indeed. Best wishes to you and all the staff.

Sincerely,

Tracy Scott

布朗先生您好：

我寫信來是要告知您，我必須婉拒貴公司提供的零售管理職缺了。如同我在面試中曾提及的，我同時也應徵了其他的工作。我這週收到來自芝加哥提供的管理職位。經過慎重的考慮之後，我決定接受那間公司的工作，因為那份工作比較符合我目前的職涯規劃。

我很感激您提供我面試及工作機會。很高興能與您會面，也因此學習到您與貴公司的發展計劃，真的是很棒的一間公司，也祝福您和所有同仁一切順心。

崔西．史考特 謹上

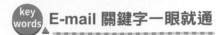

key words ▲ E-mail 關鍵字一眼就通

1 withdraw 動 取消；撤回　　2 application 名 申請　　3 indicate 動 表明

4 executive 形 主管級的　　5 career 名 職涯　　6 project 名 企劃

✈ 實力大補帖 Let's learn more!

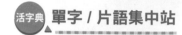

活字典 ▲ 單字 / 片語集中站

📍 撰寫婉拒信函

professional 專業的	courteous 有禮貌的
decline 拒絕	turn down 拒絕
suitable 合適的	supply sb. with sth. 提供某人…

深入學 ▲ 資訊深度追蹤

📍 婉拒工作的理由

　　在撰寫婉拒信件的時候，許多人都會卡在第二段的「婉拒理由」上。這一段往往是最難下筆、也最重要的內容。提供理由時，除了避開批評對方公司的言論外，最重要的是「態度誠懇」，以下提供幾個可以思考的婉拒方向：

01 外部因素：舉凡公司地點、上班時間、到家庭因素，都可以歸納在這一項中。

02 資歷或背景不符：在面試後，若認為工作的內容與自己的資歷不符，擔心自己無法勝任該項工作，也可以於信中說明。

03 薪資：這一點許多人都會避開，但其實只要你態度有禮、誠懇，並說明自己有必須考量薪資的難處，也是可以寫出來，並請對方體諒的。

04 職涯規劃：如果對於職涯有明確的目標，而該工作並不符合自己的規劃，也可以寫出來，但要注意的是，用字不要強調「對方公司不符合自己所想」，而應該著墨於「因為職涯規劃的關係，所以你目前需要尋找另一個領域的工作經驗」。

6-04 拒絕工作信函 II

　　本封拒絕工作信函為 Rejection Letter。Rejection Letter 與 Withdrawal Letter 的不同之處在於，Withdrawal Letter 是指「當 A 公司已經有意錄取你，但同時你又接到 B 公司提供更適合的職位時，回覆給 A 公司的拒絕信函」。withdrawal 有撤銷、撤回的意思；而 rejection 則表示拒絕。

From:	Mark Gerard
To:	Ms. Anderson
Subject:	Mark Gerard – Rejection Letter

Dear Ms. Anderson:

Thank you very much for offering me the position of retail management in your company. I appreciate your discussing the details of the position with me and giving me time to consider your offer.

The organization is impressing, and there are many aspects of the position that are very appealing to me. However, I believe it will benefit both of us best that I decline your kind offer. This has been a difficult decision for me, but I believe this is what I have to do at this point of my career.

I want to thank you again for the consideration and the kindness given to me. It was a pleasure meeting you and your outstanding staff. Wish you all the best.

Sincerely,

Mark Gerard

安德森女士您好：

謝謝您提供我貴公司零售管理的職位。感謝您耐心地向我解說這份工作的內容，也謝謝您給我時間考慮。

貴機構給人的印象真的很好，這份工作的許多層面也都很吸引我。然而，我相信拒絕接

受這份工作對我們雙方來說都是件好事。我考慮了很久，真的很難抉擇，但我相信這個決定對我的職涯規劃是必要的。

感謝您們的考慮及好意。能與您和您的職員見面是件開心的事，祝您未來一切順利。

馬克‧傑拉德 敬上

複製/貼上萬用句 Copy & Paste

I appreciate the offer of a secretary position at your company.	感謝貴公司提供我祕書一職。
I have received your letter of appointment for the post of Account Manager and I am much obliged.	我已收到您寄來關於客戶服務經理職位的信函，十分感激。
The benefits and salary are among the best I've ever been offered.	這家公司的福利和薪資是我遇過最好的。
I attended the interview in your company with great anticipation to be working in such a renowned company.	我滿懷期望到貴公司面試，渴望能在這樣享負盛名的公司工作。
However, I am really sorry to say that I will not be able to join your company for now.	然而，深感抱歉的是，我無法到貴公司上班。
However, as we discussed, the job requires me to commute for more than two hours a day.	然而就如我們討論過的，這份工作迫使我每天必須花兩個小時以上的時間通勤。
After careful thought, I have decided to take a position as a secretary in another company.	經慎重考慮後，我決定到其他公司擔任祕書。
Frankly speaking, I have been thinking that the position is not right for me after the interview.	坦白說，面試後我一直認為這份職缺不適合我。

I think that position suits my education background better.	我覺得那份工作和我的教育背景更加契合。
Although this job is less attractive than the one you offered, it's located much closer to my home and family.	雖然這份工作無法與您提供的工作相比，但它離我家人較近。
I hope you will understand my situation and please accept my apology.	希望您能體諒我的處境，並接受我的道歉。
I appreciate the time you spent with me.	感謝您撥出寶貴的時間給我。
Thank you once again for considering me for this post in your well-established firm.	再次感謝您這般信譽卓著的公司提供給我的職位。
I know there are many good candidates who can fill this position.	我相信有許多優秀的應徵者適合這份工作。
Hope I will have your blessing.	希望您可以祝福我。

實力大補帖 Let's learn more!

深入學 資訊深度追蹤

📍 拒絕工作信函的架構

一般而言，「拒絕工作的信函」可簡短地分為三大段：

01 1st paragraph: Acknowledge the offer.
第一段：感謝公司的錄取。

02 2nd paragraph: Show thoughtful consideration.
第二段：提出深思熟慮後之原因。

03 3rd paragraph: Express appreciation.
第三段：感謝對方。

筆記頁

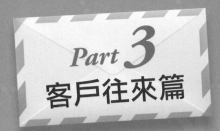

Part 3
客戶往來篇

與廠商的專業互動

Unit 1　產品推銷信

Unit 2　建立商務關係

Unit 3　產品的詢問與回覆

Unit 4　報價相關

Unit 5　訂購及回覆

Unit 6　信用交易

Unit 7　出貨、商品抱怨及致歉

Unit 8　付款與催款

 本章焦點 Focus！

✔ 推銷產品	✔ 詢價與報價	✔ 出貨與付款

　　本章的重點在於「與廠商的互動」。從初步建立商務關係開始，到購買商品，以及售後服務等，一步步教你寫出口吻適當、態度堅定的商業 E-mail。不僅僅是寫出內容，還能用文字表達出專業的職人形象。

產品推銷信
Promoting a Product

1-01 推薦新產品

　　「推銷信」的主要目的在於提供產品資訊，以引發客戶對產品的興趣。下列範例屬於簡易版的推銷信，代表撰信者與客戶的往來頻繁；反之，若收信者為新開發之客戶，則需在信件開頭簡短且清楚地介紹自己的公司與產品。

From:	Sally Wang
To:	ABC Corp.
Subject:	Recommend New Products

Dear Sirs:

We have recently designed a new product, which is selling very well in our market in Taiwan, and we are sure you will be interested in it.

We are confident that there is a potential sales prospect in your market as well; therefore, we are pleased to provide you with some samples as well as the price list.

We very much look forward to hearing from you.

Yours sincerely,

Sally Wang

Sales Director, Imodel Corp.
+886-2-3344-6688

敬啟者：

我們最近設計了一項新產品，在台灣市場的銷售表現十分亮眼，相信您會對它有興趣。

對於本產品在您們市場的銷售潛力，我們非常有信心；因此，我們很樂意提供樣品以及價目表。

我們期望能夠得到您的回覆。

王莎莉 謹上

愛模股份有限公司

業務處長
+886-2-3344-6688

複製 / 貼上萬用句 Copy & Paste

通知新品發表會

We take this opportunity to inform you.	我們趁此機會通知您。
Please allow us to call your attention for our new product launch event.	請容許我們告知您新品發表會的消息。
May we ask for your attention to our new product launch event.	我們想要告知您新產品發表會的消息。
In reply to your letter of the 5th of September, I'd like to inform you that our new product has already launched in market.	回覆您九月五日的來信，我想要通知您：我們的新產品已經上市了。
In the field of shoes market, our products have long been considered as the best among other brands.	長久以來，我們的產品在鞋業市場被視為最優秀的。
Market competition has not affected the quality and the appealing factors of our products.	同業競爭並未影響我們的產品品質以及吸引力。

As the information you have received indicated, the XYZ Motor Show will take place in October.

如同您已收到的資訊，XYZ車展將於十月舉辦。

We hope to have you here to take a look at our latest vehicles.

希望您能前來鑑賞最新車款。

We have heard that your company is seeking for kitchenware.

我們聽聞貴公司正在尋找廚房用品。

回覆詢問 & 提及人脈

Thank you very much for your inquiry about our products on the 20th of April.

非常感謝您四月二十日所寄來的詢問信函。

We are happy to inform you that we have just launched our new products.

我們很開心地通知您：我們剛推出新產品。

The object of this letter is to tell you that we have developed a new product.

此函通知您：我們已經研發了新產品。

It has better functions than the older generations.

其功能比其他舊式的產品好。

Your name has been brought up by your friend, Mr. Marvin Chen, who advised us that you are looking for a computer devices manufacturer.

您的朋友陳馬文先生提供您的名字給我們，表示您正在尋找電腦設備的製造商。

Mr. Walker indicated that you may be interested in our products.

沃克先生表示您可能會對我們的產品感興趣。

強調公司優勢

We manufacture a wide range of products.

我們製造的產品品項相當廣泛。

We produce many different products.

我們生產多種商品。

We have enjoyed an excellent reputation for quality products for over 30 years.

我們的產品已享有超過三十年的優良信譽。

We are certain that our products will satisfy your customers.

我們相信您的客戶會滿意我方的產品。

We work in close cooperation with many local renowned manufacturers.

我們與本地許多知名製造商互有往來。

We are pleased to offer you the lowest possible prices.

我們很開心能提供您最優惠的價格。

We are enjoying a good reputation all over the world.

我們在世界各地享有優良商譽。

We believe that market demand will be increasing in the future.

我們相信未來的市場需求會增加。

宣傳產品

We are the manufacturer of vehicle key parts.

我們是汽車零件製造商。

Our products are easy to operate.

我們的產品很容易操作。

Our new products are easy to use.

我們的新產品很容易使用。

One of our main advantages is its fashionable design.

我們最大的優勢在於產品的時尚設計。

Our new products are more competitive in both the quality and price.

我們的新產品在品質以及價格上皆更具競爭力。

I assure that your customers will be interested in our new products.

我保證您的客戶會對我方的新產品感興趣。

This product should be attractive to customers in your market.

本產品在您的市場應該會很受歡迎。

期盼對方回覆

We await good news with patience.	我們靜候佳音。
We trust that you will reply us immediately.	我們相信您會盡快回覆。
Please let us know your decision by replying the mail.	希望您能回信告知我方貴公司的決定。
We await your reply by e-mail.	期待您的來信回覆。
We are looking forward to getting your reply by e-mail.	我們非常期待能收到您的回覆。

信件結語：提升合作的可能性

I will provide you with more detailed information in my next visit.	我下次拜訪時將提供您更詳細的資料。
Please accept our appreciation in advance for your kind attention.	請接受我們先行感謝您的關注。
We thank you for the special care you have given to the matter.	謝謝您對此事的關注。
We thank you in advance for the anticipated favor.	我們感謝您的支持。
It's our pleasure to offer you our best services.	為您提供最好的服務是我們的榮幸。
We assure you of our best services at all times.	我們保證，您隨時都能得到我們最優質的服務。
We solicit a continuance of your confidence and support.	懇請您持續的信賴與支持。
We hope to receive a continuance of your kind patronage.	期待您的持續惠顧。

1-02 推薦智慧型手機

透過強調新產品的優點、提供試用樣品或購買優惠，可刺激收件者購買該產品的欲望。但切記：勿誇大產品的功能，以免引來損害商譽的危險；另外，也要記得留下聯絡方式，以方便對產品有興趣的收件者隨時與你聯絡。

From:	Tommy Jackson
To:	all@mail.net
Subject:	Recommend 2018 New Smartphones

Dear Customers:

We are pleased to send you our 2018 Smartphone **Brochure**[1]. We hope you find it interesting.

We have **manufactured**[2] the latest smartphone, which can offer more advanced computing ability and can be used as a **handheld**[3] computer **integrated**[4] with a mobile telephone. They are **fashionable**[5] in appearance, and with **appealing**[6] colors. The touch **panel**[7] is 2.6 inch, with large memory storage. This new smartphone supports various media format, including MP3 and MP4. It also has Bluetooth feature. It is a **combination**[8] of a digital camera and a personal digital assistant.

Judging from the undeniable increased demand for smartphones, we believe it will be a smart move to **invest**[9] in this market full of potential. Please let us know if you wish to receive details on any particular types. It is our pleasure to send you more information.

With over 10 years of experience in the smartphone market, we believe that our products will be satisfactory to both you and your customers.

Hope to have more opportunities to cooperate in the future.

Very truly yours,

Tommy Jackson

親愛的顧客您好：

很高興能寄送二〇一八年的手機目錄給您，希望您對產品感興趣。

我們已推出最新款的智慧型手機，擁有更先進的電腦功能，如同將手提電腦整合至手機中。除此之外，還擁有時尚的外觀與搶眼的顏色。本機型的觸控面板為二點六英吋，並擁有超大記憶容量。此外，還支援各式媒體播放器（包括 MP3 以及 MP4），同時也支援藍芽模式。它完全整合了數位相機與個人數位助理的功能。

由於對智慧型手機的需求日增，我方建議您投資這塊潛在市場。若想進一步了解產品，請告知我們，我們很樂意提供更多資訊。

我們擁有十年以上的市場經驗，相信您與您的客戶都會滿意我方的產品。

希望未來我們可以有更多機會合作。

湯米・傑克森 敬上

key words ▲ E-mail 關鍵字一眼就通

1 brochure 名 小冊子　　2 manufacture 動 製造　　3 handheld 形 手持的

4 integrate 動 使結合　　5 fashionable 形 流行的　　6 appealing 形 有魅力的

7 panel 名 控制板；面板　8 combination 名 結合　　9 invest 動 投資

複製 / 貼上萬用句 Copy & Paste

提供樣品或試用

We are glad to provide you with our samples.	我們很樂意提供樣品給您。
We would like to send you a sample and the user manual.	我們會寄發樣品以及說明書給您。
If you are interested in our products, please let us know.	如果您對我們的產品有興趣，請告訴我們。

Please contact us if you are satisfied with our samples.	如果您滿意我們的樣品，請與我們聯絡。
The products are now in stock.	目前產品皆有現貨。
Delivery will be made within one week after receiving your order.	收到訂單後，我們將於一週內出貨。
We look forward to receiving your test order soon.	希望不久後就能收到您的試用訂單。

提供優惠

Please see the attached price list, and let us know your requirement by replying the mail.	請參考附件的價目表，並回信告知您的需求。
You may be able to get 15% discount on your initial order.	初次訂購可享八五折優惠。
You will receive a special discount of 15% off until September 30.	九月三十日前，您可享有八五折優惠。
The special discount now offered can only be applied to orders placed by the end of June.	本次優惠僅提供給六月底前下單的訂購者。

表達合作意願

We would be pleased if there is an opportunity to cooperate with you.	若有機會與您合作的話，我們將非常開心。
We hope to get into contact with you soon.	希望能夠盡快與您聯絡。
We hope that we will have the opportunity to cooperate.	希望能有合作機會。

| 🖱 You can contact us whenever you want. | 您隨時都可以聯絡我們。 |
| 🖱 You can call us at anytime. | 歡迎您隨時來電。 |

實力大補帖 Let's learn more!

活字典 單字 / 片語集中站

📍 **手機周邊硬體設備**

earphone 入耳式耳機	headphone 頭戴式耳機
phone charger 手機充電線	portable charger 行動電源
adapter 轉接頭	smartphone holder 手機架
screen protector 螢幕保護貼	smartphone case 手機保護殼
tempered glass screen protector 強化玻璃保護貼	

📍 **收訊與網路**

reception 訊號	bad/poor reception 收訊不良
data roaming 網路漫遊	prepaid card 預付卡
unlimited data plan 網路吃到飽	

📍 **拍照相關用語**

take a picture 照相	action shot 動態相片
close-up shot 近拍相片	wide shot 廣角照片
screenshot 截圖	pixel 畫素
contrast 對比度	brightness 明亮度
turn off the flash 關掉閃光燈	set the timer 設定計時器

Part
1
入門篇

Part
2
求職篇

Part
3
客戶往來篇

Part
4
人際互動篇

From:	Fred Smith
To:	Mr. Chang
Subject:	Recommend K-Series A/C

Dear Mr. Chang:

We have learned that you are going to **renovate**[1] your old building. Here we would like to introduce our new product, Zono Multizone K-Series for your reference.

We are the market leader in A/C (Air Conditioners). Our products are not just for cooling, but provide air conditioning functions. The Zono Multizone K-series provide a flexible **switch**[2] between cooling and heating; it can power from 2.5kW to 6.0kW when operate alone, and 3.5kW to 14.4kW when separate.

The outdoor device can support up to 6 indoor devices, whether wall-**mounted**[3] or floor-mounted. And this product is available in white, silver, and **beige**[4] with dark brown colors. Also, a washable **carbon**[5] and **antibacterial**[6] air **purifying**[7] filter ensures that the micro dust, **pollen**[8] **particles**[9] and odors that can collect on filters can be easily removed. The filters can be washed and re-used up to 20 times.

The products are reasonably priced and easy to install. It also helps solve problems of odor and smoke. The system is suitable not only for households and offices, but also a good choice for the entire office building.

We will send you some other brochures for more information. We would appreciate your consideration. If you have any further questions, please contact us.

Very truly yours,

Fred Smith

張先生您好：

我們聽說您有改建大樓的需求，因此想要向您介紹我們的新產品：梭農多區域 K 系列。

我們的空調系統在市場上領先群雄，不僅只於冷卻空氣，更提供空氣調節的功能。梭農多區域 K 系列可提供冷熱調節，單獨運轉時的冷氣能力為二點五至六千瓦，分開時則可為三點五至十四點四千瓦。

一部戶外機可控制多達六台的室內機（壁掛式及鑲嵌式皆包含在內）。目前提供白色、銀色、米色三款搭配深咖啡色。此外，可清洗的活性碳除菌濾網能確保遭濾網阻絕的微塵、花粉微粒以及異味被輕易洗除。濾網經清洗後，可重複使用多達二十次。

本產品的價格合理且安裝容易，同時可解決異味與菸味的問題。不僅適用於住家與辦公室，也很適合安裝於整棟式的辦公大樓。

我們會寄幾本手冊給您，並感謝您的考慮。如果您有任何疑問， 請聯絡我們。

弗瑞德·史密斯 敬上

key words ▲ E-mail 關鍵字一眼就通

1 renovate 動 修繕
2 switch 名 轉換
3 mounted 形 鑲嵌的
4 beige 名 米色
5 carbon 名 碳
6 antibacterial 形 抗菌的
7 purify 動 淨化
8 pollen 名 花粉
9 particle 名 微粒

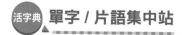

實力大補帖 Let's learn more!

活字典 ▲ 單字 / 片語集中站

📍 省電相關單字

energy 能源	consumption 消耗量
temperature 溫度	ceiling fan 吊扇
programmable 程式控制的	thermostat 自動調溫器
maintenance 保養維護	turn off 關閉（電器）

Part
1
入門篇

Part
2
求職篇

Part
3
客戶往來篇

Part
4
人際互動篇

1-04 推薦隱形護膜

From: Joan Lin

To: Mr. Hunter

Subject: Recommend Opsite Spray

Dear Mr. Hunter:

We're pleased to inform you that with years of research, we've successfully introduced our new products Opsite **Spray**[1] to the market.

What makes this product a hot sale item is its water-**resistant**[2] feature, allowing it to be applied on any types of **wounds**[3]. It is easy to apply, waterproof, ultrathin and breathable, and particularly useful in awkward areas.

This product is developed and brought to the market in response to a long-existing demand. It's now the **takeoff**[4] stage for this product, and the sales **volume**[5] increases rapidly.

Considering that you might be interested, we've sent you some catalogues and samples for you to test on. You'll find the product **extraordinary**[6] useful after trying it.

Sincerely yours,

Joan Lin

親愛的杭特先生：

很開心通知您：經過多年研發，我們的新產品「隱形護膜」已推出上市。

本產品之所以能熱賣，是由於它的防水能力出眾，受傷時可塗抹於任何患部。好塗、防水、超薄、透氣，特別適用於難以處理的傷口。

本產品的上市是為了因應市場上早已存在的需求，商品目前正在起步階段，而銷售量已經在急速增加中。

考慮到您或許會有興趣，因此寄出目錄與樣品給您，相信您試用過後，會發現此產品的非凡功效。

林瓊安 敬上

 E-mail 關鍵字一眼就通

1 spray 名 噴液

2 resistant 形 抵抗的

3 wound 名 傷口

4 takeoff 名 開始；起飛

5 volume 名 量；額

6 extraordinary 形 非凡的

實力大補帖 Let's learn more!

 單字 / 片語集中站

📍 急救箱常見用品

antiseptic 消毒水	bandage 繃帶
band-aid OK 繃	cotton swab 棉花棒
eye drops 眼藥水	gauze （醫用）紗布
Neosporin 萬用軟膏	saline solution 生理食鹽水
povidone-iodine solution 優碘	thermometer 體溫計

資訊深度追蹤

📍 家用醫藥箱的存放法

01 避免置於高溫處：以免化學藥物因為溫度而變質。

02 放置於幼兒無法取得的地方：盡量置於高處，以免幼兒誤食。

03 建議保留原藥品的標籤：不要丟棄外包裝與說明書，如果藥品沒有標籤，也要存放在乾淨的容器中，並標明藥品名稱、用法及用量、保存期限等資訊。

04 定期檢查藥物的保存期限：過期的藥物往往會變質，絕對不要因節省而留下，變質的藥物更是要立即丟棄（無論有沒有超過保存期限）。

Part
1
入門篇

Part
2
求職篇

Part
3
客戶往來篇

Part
4
人際互動篇

1-05 推薦信用卡

From: taiwanbank.Helen@mail.net

To: jack1551@mail.net

Subject: Taiwan Bank Credit Card

Dear Sir:

If you are seeking to apply for a **credit card**[1], Taiwan Bank's Credit Card is just the right one for you!

Taiwan Bank's core value is "customer first". We have extremely low **interest rates**[2], especially for first-time card holders like you!

Moreover, you can apply for our credit card online, and will be automatically **signed up**[3] to our online banking system without much paperwork.

After becoming a member, you'll enjoy a 5% reward point on each season, which can be applied in several **grocery**[4] stores, restaurants, department stores and even home improvement stores. It's free and easy to **activate**[5] your bonus each **quarter**[6].

You can also call our customer service line at 2258-3333 to get more information now!

We can't wait to hear from you!

Sincerely yours,

Helen Chen

先生您好：

若您正在尋找信用卡，台灣銀行信用卡將會是您的首選！

敝行一向秉持「客戶至上」的原則。我們的利息很低，尤其是對於像您這樣初次申辦的客戶而言更是如此。

您也可以利用線上申請的服務，申辦我們的信用卡。網路銀行的系統將自動幫助您申請註冊，免去不必要的文書作業。

加入會員之後，您每季可享有百分之五的紅利回饋，這些紅利將能於各大雜貨超商、餐廳、百貨及家用品商店使用。這項服務是免費的，而且使用紅利的方式也非常簡單。

您也可以撥打我們的服務專線：2258 -3333，以便得到更多資訊。

希望能夠得到您的回覆！

陳海倫 敬上

key words ▲ E-mail 關鍵字一眼就通

1 **credit card** 片 信用卡   2 **interest rate** 片 利率   3 **sign up** 片 註冊

4 **grocery** 名 食品雜貨   5 **activate** 動 使活動起來   6 **quarter** 名 付款的季度

實力大補帖 Let's learn more!

文法王 文法 / 句型解構

📍 申辦實用句

1 **I'd like to apply for a credit card.** 我想要辦信用卡

2 **What documents should I bring in?** 我要帶哪些文件呢？

3 **Do you have a chop and two pieces of ID with you?** 您有帶印章和雙證件嗎？

4 **Here are my passport and the necessary information.** 這是我的護照和所需資料。

5 **I'd like to apply for a balance statement of my account.** 我要申請存款證明。

6 **You have exceeded your credit limit.** 您已經超出您的信用額度了。

1-06 推薦辦公室用品

From: doris.w@mail.net

To: adam01234@mail.net

Subject: Bentley Office Supplies

Dear Sir:

Have you ever noticed how much a company spends on office supplies? If you think about it, the amount that spent on office supplies such as pencils, post-it notes, writing pads, etc. are very large! And this is how Bentley Office Supply can assist you.

Make us your one-stop office supply store! We can take your orders by phone, fax, or e-mail, and we can deliver your order within 24 hours. We will also offer you a 5% discount on each purchase, which will save you a lot!

I have attached our company catalogue and price list. I'll be calling you and hope we can set an appointment next week. We at Bentley are eager to serve you.

Very truly yours,

Doris Wang

先生您好：

關於公司的「辦公室用品」，您是否注意過其消耗量呢？若您加以計算的話，會發現花在紙筆、便利貼、以及便條紙等文具用品上的開銷其實很大！這是我們班特利文具公司可以協助您的地方。

讓我們成為您一次購足辦公室用品的供應商！我方接受電話、傳真或電子郵件訂單，二十四小時內即可送貨，並提供九五折優惠，讓您省更多！

附件為產品目錄與價目表。我會致電給您，希望下週能安排會面。我們渴望為您服務。

朵莉絲‧王 敬上

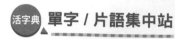

實力大補帖 Let's learn more!

活字典 單字 / 片語集中站

📍 寫字與繪畫用具

ballpoint pen 原子筆	color pen 彩色筆
crayon 蠟筆	highlighter 螢光筆
fountain pen 鋼筆	marker 麥克筆
pencil 鉛筆	eraser 橡皮擦
correction pen 修正液	correction tape 立可帶
ruler 直尺	compasses 圓規

📍 其他文具用品

stationery 文具	file folder 資料夾
post-it pad 便利貼	glue 膠水
glue stick 口紅膠	magic tape 隱形膠帶
duct tape 大力膠帶；萬用膠帶	binder clip 長尾夾
paper clip 迴紋針	pushpin 大頭釘
thumbtack 圖釘（圓扁頭型）	stapler 釘書機
staples 釘書針	scissors 剪刀
box cutter 美工刀	paper cutter 裁紙器
hole puncher 打洞器	pencil sharpener 削鉛筆器
set square 三角板	protractor 量角器

Unit 2

建立商務關係
Building a Business Relation

2-01 出口商 / 賣方尋求合作夥伴

　　想要尋找買方,並與其建立合作關係時,可簡短地介紹自己如何得知對方的訊息,並說明希望與買方往來的理由以及合作之後的好處;此外,建議附上商品目錄、價目表與公司的聯絡資訊,以便進一步的業務往來。

From:	Paul Jenkins
To:	B&I Corp.
Subject:	Establish Business Relations

Dear Sir or Madam:

We are a manufacturer and **exporter**[1] of **industrial**[2] **chemical**[3] **raw**[4] materials from England. We are expanding our target markets this year, and would like to find a business partner. If you are interested in building business relations, please visit our website at: http://www.chemicalexporting.com.

Please feel free to contact us and discuss more about your company background, and how we can assist and support you in order for you to make more **profit**[5] by exporting the industrial chemical raw materials from England.

Enclosed please find a copy of our **illustrated**[6] catalogue, covering the main products available at present. We look forward to your reply about future cooperation.

Sincerely yours,

Paul Jenkins
Sales Manager

敬啟者：

我們是英國工業化學原料製造暨出口商。我們從今年開始擴大市場，所以在尋找合作夥伴。若您對敝公司有興趣的話，請參考下面的網頁：
https://www.chemicalexporting.com

歡迎您與我們討論貴公司的背景，以及我們能提供的資源，希望能藉由我們從英國出口化學原物料，幫助貴公司取得更多利潤。

隨信附上商品目錄，裡面包含我們目前的主要商品，敬請告知貴公司的合作意願。

業務經理　保羅・詹金斯 敬上

 E-mail 關鍵字一眼就通

1 exporter 名 出口商　　2 industrial 形 工業的　　3 chemical 形 化學的

4 raw 形 未加工的　　5 profit 名 利潤　　6 illustrate 動 說明

 複製 / 貼上萬用句 Copy & Paste

得知對方公司資訊的渠道

Your name and address were listed in The Chemicals magazine.	貴公司的名稱與地址刊登在《化學雜誌》上。
Your company was introduced to us by Dr. Harris.	我們是透過哈里斯博士的介紹，得知貴公司的資訊。
Your company was highly recommended to us by the Trade Center.	貿易中心向我們強力推薦貴公司。
We learned from the Internet that your company specializes in jewelry.	我們從網路上得知貴公司從事珠寶業。
We received your information from the British Trade and Cultural Office.	我們從英國經貿文化辦事處得知貴公司的資訊。

表明意圖

We are writing to you in the hope of business relationship with you.	我們特寫此信，尋求與您的商務合作關係。
We are writing in the hope of opening an account with your firm.	我們特寫此信，尋求與貴公司的合作關係。
We are now seeking cooperation opportunities.	我們正在尋求合作機會。
Our company is seeking cooperation opportunities.	敝公司正在尋找合作機會。
We wish to expand the business of our earphones to your country.	我們希望能夠在您的國家擴大耳機業務。
We are looking for partners to expand the market in Europe.	我們正在尋找歐洲市場的合作夥伴。
We are keen to establish a business relationship with you.	我們很希望能夠與貴公司建立商務關係。
We are desirous to build connection with the most reliable company in your country.	我們想要與貴國最值得信賴的公司建立商務關係。
We would like to find a company who would represent us.	我們希望能與願意做我方代表的公司搭上線。
We are desirous of expanding our services to Asia.	我們渴望在亞洲擴展我們的業務市場。
We are desirous of enlarging our trade business in Japan.	我們想要在日本擴展業務。
We are expanding our market and therefore, we hope to find a partner.	我們正在拓展市場，希望找到一位合適的夥伴。

介紹自己公司

Please allow me to introduce our company to you.	請容許我向您介紹敝公司。
Our company specializes in electric fans.	敝公司主要生產電風扇。
We are one of the leading exporters of perfume.	我們為知名的香水出口商。
We are a newly established company, based in Tainan, offering travel bags.	我們是一家剛於台南成立的新公司，主要供應旅行袋。
We have earned a good reputation in both Asia and Europe.	我們在亞洲以及歐洲都擁有良好的聲譽。
Our products are very popular both locally and internationally.	我們的產品在本地以及國外都非常受歡迎。
Our goods are already being sold in many countries in Asia.	我們的產品已在亞洲多國販售。
Now we are exporting our goods to many countries in Europe, and are also trying to increase some Asian buyers.	我們目前主要將貨品出口到歐洲國家，也正積極尋找亞洲買家。
We have an excellent reputation through 20 years of trade experience.	我們已有超過二十年的貿易經驗以及優良聲譽。
We are sure you will be satisfied with our products and the excellent quality.	我們確信您會滿意我們所提供的高品質商品。
We are sure that our products would be of great interest to you.	我們確信您會對我們的產品感興趣。

積極地表態

We will give you our lowest quotations and try our best to comply with your requirements.	我們可以因應您的需求，給您最低報價。
You are warmly welcomed to visit our factory and to discuss business with us.	我們誠摯地歡迎您來我們的工廠參觀，並一同商討貿易關係。
We hope that our products will be added to your sample categories.	希望我們的產品能新增到您的樣品目錄上。
We would like to be your supplier in Taiwan.	我們希望能成為您在台灣的供應商。
Please see the attached files for our company information.	附件為敝公司資訊。
If you would like to know more about our company and products, please contact us.	若您想進一步了解敝公司，請與我們聯繫。
If you are interested in cooperation, please feel free to contact us.	若您有合作的意願，請與我們聯繫。
If you have further queries, please feel free to contact me at 0800-112-233.	如果您還有其他疑問，請致電給我，號碼為 0800-112-233。
Thank you for your attention.	謝謝您的關注。

進口商 / 買方尋求合作夥伴

在尋找賣方，與對方建立合作關係時，要簡短說明如何得知賣方之產品訊息與其產品優勢，並清楚告知我方的行銷能力以及能夠提供給賣方的服務，以尋求進一步合作。

From:	Sandra Wang
To:	XYZ Corp.
Subject:	Establish Business Relations

Dear Sirs:

A few days ago, we had the opportunity to see a **showcase**[1] of your products at the New York Fashion Show. We were very **impressed**[2] with the **quality**[3] and the **affordable**[4] prices.

We would like to introduce a trade service to you. We have excellent references and **connections**[5] in the trade, and are fully experienced with the import business for this type of product.

We are sure that we can sell a great amount of your products, if you would allow us to **promote**[6] sales in Taiwan.

Please let me know what you think.

Respectfully,

Sandra Wang

敬啟者：

幾天前，我們在紐約時尚展看到貴公司的產品，我們對於其品質和價位印象深刻。

我們想要提供敝公司的貿易服務。在此產品業界，我們的推薦客戶和往來公司都很廣，也擁有完整的進口經驗。

若您能讓我們在台灣促銷，我們有信心能促成大量的銷售成績。

煩請給我方建議回函。

珊卓拉・王　敬上

key words ▲ E-mail 關鍵字一眼就通

1 showcase 名 陳列　　2 impress 動 使印象深刻　　3 quality 名 品質

4 affordable 形 負擔得起的　　5 connection 名 業務往來　　6 promote 動 宣傳

複製 / 貼上萬用句 Copy & Paste

得知對方公司的渠道

Your information was given to us by the Taipei Trade Office.	我們透過台北貿易中心得知您的資訊。
We acquired your name and address from your advertisement in Fashion Look magazine.	我們從《時尚觀點》雜誌得知您的公司名稱與地址。
We saw your advertisement in Fashion Look magazine this Tuesday.	我們本週二在《時尚觀點》雜誌上看到您的廣告。
We acquired your information from our friend - Ms. Chang of Wonderland Trading Company.	我們從旺德蘭貿易公司的張小姐那裡得知您的資訊。
Your company was referred to me as a potential supplier.	您的公司被推薦為潛在供應商。
Your company was recommended by Taiwan Trade Center in France.	您的公司為駐法台灣貿易中心所推薦。

推薦自己的公司

We take the chance to introduce you our company.	冒昧地向您介紹敝公司。
Our company is a leading importer of quality men's wear.	我們是知名的男性服飾進口商。

We are an importer of children's shoes.	本公司的主要業務為進口童鞋。
We have many shops in Taiwan.	我們在台灣有許多經銷據點。
We can assure that your turnover will increase considerably.	我們向您保證：您的產品營業額將會有大幅度的成長。
Our company is very well connected with many jewelry dealers in Taiwan.	本公司與台灣許多珠寶經銷商均建立良好的合作關係。
We are sure we can sell large quantities of your products.	我們確信能大量銷售貴公司的產品。
There is a large demand for various kinds of cosmetics in Taiwan.	多樣化的化妝品在台灣的需求量非常大。
We have many representatives in all major cities of Taiwan.	我們在台灣的幾個主要城市都有經銷業務。
We have considerable experiences in this field.	在這塊領域，我們的業務經驗非常豐富。

表達需求 & 合作意願

We are seeking new suppliers for good quality women's apparels.	我們正在尋找品質良好的女性服飾供應商。
We are looking for a partner to supply us with such products.	我們正在尋求提供此產品的供應商。
We learned that you have been supplying the best quality furniture to all over the world.	我們得知貴公司生產高品質的傢俱，並銷往世界各地。
Some of your goods look very interesting.	貴公司的某些產品看起來非常不錯。
We are very interested in knowing more about your products.	我們想要進一步了解貴公司的產品。

Please send us your brochure and catalogue of all your men's wear.	請寄貴公司的男性服飾型錄給我們。
Please also include the price list.	也請附上價目表。
Please indicate your requirements on the terms of payment, delivery, and minimum order quantity.	請告知您的付款條件、運送方式以及最低訂單量。
If samples are available, we would like to take some for our reference.	如果能夠提供樣品，我們希望能夠參考一下。
If we are satisfied with the samples, we would place an order soon.	若樣品令人滿意的話，我們就會下訂單。

信件結語

Would you please inform us if you have any partners in Taiwan?	可否告知您在台灣是否有其他合作的公司呢？
We hope we can cooperate soon in the near future.	我們希望在不久的將來能與您合作。
We would appreciate your quick response.	若您能盡速回覆的話，我們將會非常感謝。

實力大補帖 Let's learn more!

文法王 文法 / 句型解構

表達合作意願

1 sb. had the pleasure of V-ing 某人很高興能…。

2 sb. would appreciate/be obliged receiving... 某人將很高興能收到…。

3 If...live up to our expectations, we would like to order regularly. 若…符合我們的期望，我們將會長期訂購。

替自家產品尋找代理商

　　尋求代理商服務時，可概略說明如何得知此公司，並簡短地分析自家產品的銷售潛力，並說明與對方（代理商）的結算方式，以供對方參考。

From:	Fiona Johnson
To:	tradingex@mail.net
Subject:	Searching for Exclusive Agent in China

Dear Sir or Madam:

We are a manufacturer of **keyboards**[1], and we are seeking a **representative**[2] in China. As you are one of the leading sales agents in your country and full of potential, we would like to know if you are interested in being our **sole**[3] agent in China market.

As you will see from the enclosed catalogue, our products are popular in many countries. With **competitive**[4] prices and quality, we have decided to **expand**[5] our market to China. We will offer a 10% **commission**[6] on the list prices for our sales agent.

If you are interested in developing this market, please write us back as soon as possible.

Sincerely yours,

Fiona Johnson

敬啟者：

我們是電腦鍵盤的製造商，正在尋找中國區代理商。因為您在貴國為極有潛力的代理商，因此我們來函詢問，想了解貴公司是否願意成為我方在中國的總代理。

附件為產品目錄，我們的產品已在多國銷售，很受歡迎，因為價格與品質皆具競爭力，因此決定拓展中國市場。若您願意成為代理商，我方將提供百分之十的佣金。

若您有興趣開發此市場，請儘速來信告知。

費歐娜‧強森 敬上

E-mail 關鍵字一眼就通

1 keyboard 名 鍵盤　　2 representative 名 代表　　3 sole 形 唯一的

4 competitive 形 競爭的　　5 expand 動 擴大　　6 commission 名 佣金

✈ 複製 / 貼上萬用句　Copy & Paste

信件開頭 & 自我介紹

We get the information of your company from Taipei Chamber of Commerce and hereby writing to you for more details.	我們從台北商會那裡得知貴公司，特此來函。
You were recommended to us by Taichung World Trade Center.	台中世界貿易中心向我們推薦貴公司。
We are one of the main producers of scanner.	我們是掃瞄機的主要製造商之一。
We are the leading exporter of computer supplies.	我們為知名電腦週邊產品的出口商。

展現自身優點

We can offer a wide range of products.	我們能提供各式產品。
Our products have attractive design and a one-year warranty.	我們產品的設計新穎，且有一年保固。
We are a manufacturer of high quality mouse for computers.	我們為高品質滑鼠的製造商。
We already exported our products to many European countries.	我們已出口商品到歐洲各國。
We have full confidence in our products.	我們對自己的產品有信心。

提供對方誘因

We believe that there is an enormous potential market to be developed in China.	我們相信中國是個極具開發潛力的市場。
This would be a great opportunity for you to grow your target markets.	這對貴公司來說將會是一個擴展市場的好機會。
It's an excellent opportunity to establish a wider range of customers.	這是一個擴展客戶群的絕佳機會。
We will offer 10% commission to our agent, plus advertising funding support.	我們準備提供百分之十的佣金給代理商，並提供廣告支援。
We would supply you with our prices at 10% below the export price list.	我們願意提供低於出口報價百分之十的佣金。
We are sure this is a golden opportunity to invest in.	我們確信這是一個絕佳的投資機會。
If you are interested in our proposal, please let us know the terms for commission and other related charges.	若您對本公司的提案有興趣，請告知我們您的佣金條件以及其他費用。

談合約 & 結語

The contract will be effective from Jan. 1 for one year.	合約將自一月一日起生效，為期一年。
Under both parties' agreement, the contract will be renewed for a further year.	如經雙方同意，合約可再延長一年。
We hope to hear favorably from you soon.	我們希望能夠儘早得到您善意的回應。

Part
1
入門篇

Part
2
求職篇

Part
3
客戶往來篇

Part
4
人際互動篇

2-04 向企業提供代理服務

申請成為代理商時，記得要強調自己公司的實力以及優勢，以引起對方的注意，進而產生與你洽談代理之意願。

From:	Ito Takumi
To:	ABC Inc.
Subject:	Import Electric Watches

Dear Sirs:

Watch Tech is a reliable Japanese company with wide and varied experiences in this industry. We see great potentials in your watches to be sold in Japan. As we have a team of experienced sales representatives, we propose to act as your sole agent for electric watch business in Japan.

Enclosed is our Agency Contract in detail for your reference. We are looking forward to hearing your reply as soon as possible. Thank you in advance for your time.

Very truly yours,

Ito Takumi

敬啟者：

錶德為一家在日本信譽卓著的公司，在這個業界擁有豐富的經驗。我們相信貴公司的錶品在日本市場具有銷售潛力。我們有經驗豐富的銷售人員，現自我推薦，作為貴公司電子錶在日本的獨家代理商。

隨信附上代理權的合約內容，以供您參考。感謝您撥冗閱讀，期盼不久後就能收到您的回覆。

伊藤拓海 敬上

表達意願：提供代理服務

We learned that you haven't had an agent in Taiwan.	我們得知貴公司在台灣沒有代理商。
We have been informed that you are seeking for an agent.	我們得知您正在尋找代理商。
We are interested in your advertisement of March 13 about a Japanese local agent recruitment.	您於三月十三日所刊登，關於尋找日本區代理商的廣告引起我們的興趣。
We would like to offer our services as an importer of your products.	我們願意提供您進口產品的服務。
We would like to offer an agency service.	我們願意提供您代理商的服務。
We are interested in an exclusive arrangement with your factory for the promotion of your products in the Middle Eastern market.	我們希望能獨家代理貴公司的產品，負責中東地區市場的銷售。
The excellent quality and modern design of your products appeal us very much.	貴公司產品的優良品質以及新穎的設計吸引了我們的注意。
We would like to know if you have considered expanding the market to Southeast Asia.	冒昧詢問貴公司是否有意願拓展東南亞市場。
We would like to recommend our company as the sole agent for your products.	我們想要推薦敝公司成為您的獨家代理商。

強調專業與行銷能力

We specialize in kitchen utensils in the Northeast Asia markets.

我們在東北亞市場專營廚房用具。

We have a long-time experience in marketing.

我們擁有非常豐富的行銷經驗。

We have been importing this kind of products for more than 30 years.

我們進口此類商品已有三十幾年的時間了。

We have many offices and showrooms in the major cities of Taiwan.

我們在台灣主要的城市都設有辦公室與展示間。

We are now ready to expand our sales.

我們現在已經準備好要擴展業務了。

We are sure we would be of tremendous assistance for your company to expand the market in Taiwan.

我們相信敝公司能夠大力協助貴公司拓展在台灣的業務。

We will devote full attention to establishing your products In Southeast Asia.

我們將會全力投入在東南亞銷售您的產品。

We are confident that we will be able to provide a wide range of customers for your goods.

相信我們可以替您帶來大量的客戶。

進一步談代理

We are well prepared to do business with you if your price and terms are competitive.

如貴公司的價格以及條件具有競爭力，我們就會與您交易。

We would like to discuss the possibility of establishing agency agreement with your company.

我們希望能與貴公司討論合作代理事宜。

I will call you tomorrow to inquire about setting an appointment with you.

我會於明日致電詢問與您的會面事宜。

2-05 接受合作的提案

欲接受合作的提案時，可先於信中提出想與對方討論的項目，如貨物裝運、稅金、佣金等內容，讓對方能夠預先做準備。

From:	Tracy Miller
To:	Paul Chen
Subject:	Re: Establish Business Relations

Dear Mr. Chen:

Thank you for your mail of June 26. We are pleased to know that you find our products satisfactory. We are very interested in your **proposal**[1] and would like to expand our market to Asian Region.

We would like to discuss the **possibility**[2] of **establishing**[3] a business relationship with you and look forward to **receiving**[4] your call. We would like to discuss the details, including the **shipping**[5] arrangement, the import **duties**[6], and commission.

Enclosed pleases find the latest catalogue and the price list you requested. If you'd like to have more information, please feel free to call us at 0800-123-456.

Thank you again for your interest and attention.

Faithfully yours,

Tracy Miller

陳先生您好：

謝謝您六月二十六日的來信，很高興您對我們的產品的肯定。我們對您的提案很有興趣，也有意願擴展亞洲市場。

我們希望能與您討論商務合作的可能性，也期望能接到您的來電。我們想要針對貨品的運送、稅金以及佣金等事宜與您討論。

附上您要求的最新目錄以及價目表。如附件未能提供您所需的所有資訊,歡迎您來電:0800-123-456。

再次感謝您的關注。

崔西‧米勒 謹上

 ## E-mail 關鍵字一眼就通

1 proposal 名 提議　　2 possibility 名 可能性　　3 establish 動 建立

4 receive 動 收到　　5 shipping 名 運輸　　6 duty 名 稅

 複製 / 貼上萬用句 Copy & Paste

開頭的回覆

Thank you for your mail of July. 6, introducing your company and expressing your interests in our goods.	謝謝您七月六日來信介紹貴公司,並表達對本公司產品的興趣。
We received your mail on June 20 and would like to thank you for your brief information.	我們在六月二十日收到您的來信,謝謝您的精簡介紹。
Thank you for your mail of September 15, proposing a business relationship with us.	謝謝您九月十五日來信提案欲與本公司建立關係。

表達合作意願

We are still a newly established company but is expanding rapidly.	本公司雖然剛起步,但發展快速。
We are pleased to know that you think there is room for sales expansion in Taiwan.	我們很高興得知您認為台灣市場仍有擴展的潛力。

We are very interested in your proposal.	我們對您的提案很有興趣。
We are interested in the chance of developing our trade business.	我們對於進一步發展業務的機會很感興趣。
We are thinking if you would like to give us some ideas of the terms on which you are willing to handle our goods.	我們想要了解貴公司將給予何種條件來銷售我們的產品。

談及商品

We'd also like to introduce ourselves so that you will know more about our company and products.	為了幫助您了解我們與旗下的商品，在此向您介紹一下本公司。
We have the pleasure to send you our catalogue with details of our products.	我們很高興寄送目錄與產品資訊給您。
Regarding the products you asked for, we will send you our catalogue and price list by mail.	針對您所詢問的產品，我們將寄送產品目錄以及價目表給您。
We hope you will find the catalogue useful and interesting.	希望目錄對您有幫助。

積極的安排

We can arrange your sales manager to look around our factory.	我們可以安排貴公司的業務經理來工廠參觀。
We will contact you and arrange a meeting.	我們會與您聯絡，並安排會面。
We hope that our companies can cooperate in the future.	希望我們有機會合作。
We assure you that you will be satisfied with our products with excellent quality.	我們確信您會非常滿意我們商品的優良品質。
We are sure that you will benefit from our service.	我們確信您會從我們的服務中獲益。

2-06 接受成為代理商

接受合作代理時，可詢問適當的會面時間，以便雙方進行會談以及簽署合約。

From:	Ines Moreau
To:	Mr. Chang
Subject:	Re: Seeking the Exclusive Agent

Dear Mr. Chang:

Thank you for your mail of March 6, offering us the sole agency for your LCD products in France.

We are very interested in your proposal and are confident that we should be able to develop the market in France. We have asked our Export Manager, Mr. Batson to fly over and visit your office and factories two weeks later.

Please advise the suitable date for his visit. Mr. Batson is authorized to draw up an agency agreement before flying back to Paris.

Yours faithfully,

Ines Moreau

張先生您好：

謝謝您三月六日的來信，願意授權我們在法國獨家代理貴公司的液晶顯示產品。

我們對您的提案非常感興趣，並且有信心能在法國開發市場。我們已請出口部經理，拜特森先生兩週後拜訪貴公司工廠以及辦公室。

請再告知我們方便前往拜訪的日期。此外，拜訪貴公司期間，拜特森先生將全權負責辦理簽署合約事宜。

茵妮絲・莫羅 謹上

複製 / 貼上萬用句 Copy & Paste

開頭 & 表達意願

Thank you for proposing us as your representative in Taiwan.	謝謝您來信提案敝公司成為您在台灣的代理。
Thank you for the enclosed information about your company catalogue and financial statement.	謝謝您來信附上貴公司的產品目錄以及信用狀況。
We have great interest in your proposal.	我們對於您的提案非常有興趣。
We are fond of your proposal.	我們很喜歡您的提案。

進一步商談

Your basic terms are acceptable.	我們樂於接受您提出的條件。
Please let me know when it will be convenient to call you.	煩請告知適合致電的時間,以便聯繫。
We would like to discuss trade terms.	我們希望能夠與您討論貿易條件。
We would like to visit you for having a personal interview with your sales manager.	我們希望能夠親自拜訪貴公司的業務經理。
I look forward to meeting you.	我期待與您會面。

2-07 接受成為代理商並提供協議

接受對方提案，成為代理商的同時，可於回覆合作信中條列出期望的合約內容，供對方先行檢閱，以利下一步合約之草擬及正式簽約。

From:	Ines Moreau
To:	Mr. Chang
Subject:	Conditions of Agreement

Dear Mr. Chang:

We are pleased to act as your sole agent in France. As we have many **customers**[1] in the major cities in France, they will be great **distribution**[2] outlets for your LCD products.

Before the **contract**[3] is concluded with **signatures**[4], we would like to **confirm**[5] the following terms and conditions as the basis for a formal agreement:

1. We operate as Sole Agent for your goods for a period of two years.

2. We receive a commission of 20% on the list price.

3. All advertisement cost for the first year is to be **refunded**[6] by you.

4. All advertisement cost for the second year will be divided equally between you and us.

We look forward to your confirmation letter.

Yours faithfully,

Ines Moreau

張先生您好：

很開心成為貴公司在法國的總代理商。我們有許多客戶分布在法國的主要城市，相信他們會為您的液晶顯示產品提供銷售點。

在正式簽署合約前，我們希望確認幾項主要的協議內容：

一、兩年內，我們為貴公司產品的獨家代理商。

二、代理佣金為報價的百分之二十。

三、第一年的廣告費由貴公司負責。

四、第二年的廣告費由雙方平均分攤。

希望貴公司來信確認以上條件。

茵妮絲‧莫羅 謹上

 E-mail 關鍵字一眼就通

1 customer 名 顧客　　　**2** distribution 名 分布　　　**3** contract 名 合約

4 signature 名 簽署　　　**5** confirm 動 確認　　　**6** refund 動 退還

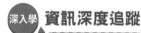 **實力大補帖 Let's learn more!**

 單字 / 片語集中站

📍 合約內容

sign a contract 簽約	come into effect 生效
reach an agreement 達成共識	terms of the contract 合約條款

深入學 資訊深度追蹤

📍 常見的合約類型

01 Contract / Agreement（合約書）：通常載明雙方的權利義務，為合約主體。

02 Memorandum of Understanding（合作備忘錄）：一般訂立於正式的買賣合約之前，藉由備忘錄建立簡單的法律關係。

03 Guaranty（保證書）/ Power of Attorney（委任書）：此類合約與英文信函相近，故又被稱為書信類合約。

2-08 婉拒對方成為代理商

收到對方的提案，若覺得不適合，就要著手寫一封拒絕信函。在此類信件中，最好提供一個合理的理由，因為日後還是可能會與對方有商業往來，所以記得口氣要委婉且有禮。

From:	anna.m@bicorp.mail.net
To:	Saito Ren
Subject:	Re: Establish Business Relations

Dear Mr. Saito:

Thank you for your mail on July 4 proposing to establish business relations between us.

Much as we are interested in doing business with you, we regret to inform that we are not in a position to enter into business relationships with any companies in Japan, because we have already had an agency arrangement with Luckwill Trading Co., Ltd. there. According to our arrangement, only through Luckwill Trading Co. can we export our products to Japan.

We regret for not being able to cooperate at this time, hope to still have cooperation opportunities in the future.

Thank you again for your proposal and understanding.

Faithfully yours,

Anna Morgan

齋藤先生您好：

感謝您於七月四日來函，欲和我方建立商務關係。

儘管我方對於和貴公司建立商貿關係非常有興趣，我們必須遺憾地通知您，由於我方已與日本拉克威爾有限公司有代理協議，因此無法和任何在貴國的公司建立商務關係。根據協議，惟有透過上述公司，我方才得以出口產品至日本。

很抱歉目前無法與貴公司合作，但希望將來仍有合作機會。

再次感謝您的提議，若您能體諒我方的立場，將不勝感激。

安娜‧摩根 敬上

複製 / 貼上萬用句 Copy & Paste

信件開頭：表達感謝

Thank you for your mail regarding representation in your country.	謝謝您來信，欲與我們討論本公司產品在貴國的代理權。
Thank you for your interests in our products.	謝謝您對我們的產品感興趣。
We appreciate your offer as our representative.	謝謝您提出「想成為本公司代理商」的提案。

拒絕提案 & 提供理由

We cannot accept your proposal.	我們無法執行您的提案。
We now have a distributer who handles our products in your country.	目前在貴國，我們已經有專門負責的代理商。
We are very satisfied with our present distributor.	我們對現在的代理商很滿意。
We have signed a contract with another company.	我們已與其他公司簽署合約了。
We regret to tell you that we made our decision prior to receiving your proposal.	很抱歉，在收到您的提案前，我們已有了其他決定。
We have no plans to make any changes currently.	我們目前尚無計畫變動。

This is not a good time to expand our business.

目前並非我們拓展業務的好時機。

It's our policy not to have any sole agents in any countries.

我們原則上不打算在任何國家設立總代理。

提供未來合作的機會

We suggest that you try to sell some of our products in your market first.

我們建議您先試著在您的市場銷售我們的產品。

We would like to establish a good business relationship with your company by selling our goods to you.

我們想藉由向貴公司銷售產品的方式，建立良好的商務關係。

Your letter has been filed for our future reference.

我們已將貴公司的信件歸檔，以備將來合作的可能性。

We will keep your name in the file.

我們會將貴公司的名單歸檔。

We will contact you at a proper time.

待時機適當，我們會與您聯繫。

If the position changes in the future, we would be in contact with you again.

如果未來情況改變，我們會再與您聯絡。

Should there be any changes, we will contact you then.

若將來情況有變，我們會再與您聯繫。

We look forward to the opportunity to discuss with you in the future.

我們希望未來能夠有機會再與您討論。

We hope our reply is not disappointing.

希望我們的答覆沒有令您失望。

Thank you for your interest in doing business with us.

謝謝您對於與本公司合作感興趣。

產品的詢問與回覆
Asking for Further Information

3-01 索取產品目錄

在索取產品目錄的信件中，首先可簡單說明來信目的，接著有禮貌地表示欲向對方索取商品型錄，並於信末表達感謝，給收件者一個好印象。

From:	Ella Chen
To:	Y&I Furniture
Subject:	Request for Catalogues

To whom it may concern,

This is Ella Chen from Zara Inc. We are soon opening our new office in Taipei and we will need some desks, chairs, and lights. Can you send us your catalogues with prices, sizes and colors for these items, please?

Thank you for your assistance.

Yours faithfully,

Ella Chen

敬啟者：

我是薩拉集團的陳艾拉，因為我們即將在台北成立新公司，所以會需要幾張桌椅和燈具。您是否能將包含價格、尺寸以及顏色等資訊的目錄寄送給我們呢？

感謝您的協助。

陳艾拉 謹上

複製／貼上萬用句 Copy & Paste

信件開頭：表明目的

We have noted your advertisement in China Post, and are interested in your products.	我們在《中國郵報》看到貴公司的廣告介紹，對您的產品很有興趣。
We are desirous of extending our connections in your country.	我們準備拓展本公司在貴國的業務。
Now we are hoping to enlarge our trade in this product.	目前敝公司亟欲擴大此項商品的業務範圍。
We're currently looking for a supplier for us.	我們目前正在尋找供應商。
Your website offers the information we need.	貴公司網站所提供的資訊符合我們的需求。
We are currently establishing our websites and are interested in some software you offer.	我們目前正在架設網頁，並對貴公司的軟體有興趣。

需要對方配合之處

Could you please send me a copy of your company's product brochure?	煩請寄送一份貴公司的產品目錄給我。
Could you please send me a copy of your latest catalogue?	請將貴公司最新的型錄寄給我。
Could you please send us the details of your current catalogue?	您是否能將最近的詳細目錄資料寄給我們呢？
I would like the information to be provided to me as paper copies.	希望能夠收到紙本資料。
We request an up-to-date catalogue and price list for your products.	希望貴公司能寄送最新的型錄及報價單。

I shall be grateful if you would send us your brochure and price list about your products.	若您能寄產品的目錄及價目表給我，將不勝感激。
Will you please send us a copy of your catalogue, with the details of prices and items of payment?	是否能寄一份貴公司的型錄，並附上價格以及付款明細呢？
I would like to be placed on your mailing list and to receive all of your catalogues for the upcoming year.	我們希望能加入您的收件者清單，並收到貴公司明年的目錄。
We should find it most helpful if you could also provide samples of these products.	如果貴公司能提供這些商品的樣品，將對敝公司有極大的幫助。
Please forward information regarding your products and services.	請提供貴公司產品及相關服務的資訊。
I am writing to request some information about the price.	我希望能夠得到產品價位的相關資訊。
We are looking forward to receiving your quotation as soon as possible.	我們期待能盡快收到貴公司的報價單。
It will be appreciated if you can also provide information for other brands so that we can make a comparison.	若能提供其他品牌的相關資訊以供我們進行比較，將非常感激。
When quoting, please state terms of payment and the time of delivery.	貴公司報價時，請說明付款條件與交貨時間。
I would also like to know the information about the delivery fee.	另外，我也想要了解貨運價位的資訊。
It would be appreciated if you also provide a comprehensive list of your charges and your contact information.	若您能提供詳細的收費表以及聯絡方式，將非常感謝。
I would also like to know if it is possible to make online purchase.	我也想知道是否可以透過網路下單。

I would also like to receive discount or special offer notices. | 我也希望能夠收到折扣與特惠的資訊。

Please also provide below information: services charges, additional software and hardware, valid warranties, and average time frames for repair and maintenance. | 也請提供以下資訊：服務的收費標準、額外的軟體以及硬體設施、有效的保證書以及平均維修週期。

期待對方回覆

Thank you for your kind assistance. | 謝謝您的協助。

We are looking forward to hearing from you soon. | 我們期待您的儘速回覆。

Your kind assistance will be appreciated, and we earnestly await your prompt reply. | 對於您的協助我們將感激不盡，敝公司將靜待您的即時回覆。

Your prompt reply will be much appreciated. | 感謝您的儘速回覆。

We look forward to receiving your reply in acknowledgement of this matter. | 我們期望能收到關於此事的確認信函。

實力大補帖 Let's learn more!

活字典 單字 / 片語集中站

目錄提供的資訊

brand 品牌	color 顏色
description 描述	image 照片
material 材質	pattern 圖案
price 價格	size 尺寸

與報價有關的單字

quotation 估價單	price list 報價單
offer sheet 報價表	reasonable 合理的
selling price 定價	margin 利潤
material cost 材料成本	labor cost 人工成本
competitive 競爭性的	rock-bottom price 最低價

文法王 文法 / 句型解構

回顧商談內容

1 As we discussed on phone this morning, ... 根據今早的電話內容，…。

2 As I mentioned last time, ... 如同上次我所提到的，…。

3 Regarding your ad. in (name), ... 根據今早您刊登在（報章雜誌）的廣告內容，…。

4 With reference to your advertisement in today's China Times, ... 根據今早您刊登在《中國時報》的廣告內容，…。

深入學 資訊深度追蹤

brochure 與 catalogue 的不同之處

在索取商品資訊時，經常會聽到這兩個單字，但其實兩者所包含的意思並不同，不能單純以頁數或紙張大小來判斷。要分辨這兩個單字的用途，可從下面兩點著手：一、用途；二、所提供的資料種類。

根據辭典，brochure 與 catalogue 的定義如下：

01 brochure：a magazine or thin book with picture that gives you information about a product or service（為雜誌或薄冊，內容提供「某項」產品、服務的資訊及相片）

02 catalogue：a catalogue is a list of things such as the goods you can buy from one company,...（目錄提供「多項」產品的資訊，比如在某家公司能購買到的全部商品）

Part
1
入門篇

Part
2
求職篇

Part
3
客戶往來篇

Part
4
人際互動篇

3-02 提供對方產品目錄

及早回信給欲索取型錄的潛在顧客是很重要的，如果可以趁機告知產品優勢或優惠更好。收到索取目錄的信函時，一定要把握機會，以免失去合作良機。

From:	Terry Chou
To:	Mr. Yang
Subject:	Re: Request for Catalogues

Dear Mr. Yang:

Thank you for your e-mail. I have sent the latest catalogue to you this morning. The catalogue **includes**[1] details of the office **equipment**[2] we **supply**[3].

You can also **view**[4] all our products on our website: www.firsthousing. com. We offer a 5% discount if you make an online **purchase**[5]. I am also **attaching**[6] a copy of our price list.

Please contact me if you have any questions or would like to have some advices.

Yours sincerely,

Terry Chou

楊先生您好：

謝謝您的來信。我今天早上已將最新的產品目錄寄給您，內含我們公司可提供的辦公室設備，並附上商品的詳細資訊。

您也可以上網瀏覽我們的網站（www.firsthousing.com），網站上列出了所有的產品。如果您透過網路下單的話，我們會提供九五折的優惠。另外，隨信附上產品的價目表給您參考。

如您有任何問題，或需要一些建議，請與我聯絡。

周泰瑞 敬上

1 include 動 包含　　2 equipment 名 設備　　3 supply 動 供應

4 view 動 查看　　5 purchase 名 購買　　6 attach 動 附加

複製 / 貼上萬用句　Copy & Paste

信件開頭

Thank you for your inquiry about our equipment.	謝謝您來信詢問我們所販售的設備。
Thank you for your e-mail on March 19.	謝謝您三月十九日的來信。
Many thanks for your message on March 19.	感謝您三月十九日的訊息。
Thank you for your enquiry.	謝謝您的詢問。
Thank you for your letter of May 21 and your interest in our translation services.	謝謝您五月二十一日來信，詢問敝公司的翻譯服務。
In response to your mail of November 21, we have enclosed our catalogue and price list.	謹依貴公司十一月二十一日的來信，附上商品目錄以及價目表給您參考。
As requested, we have enclosed the price list of our services.	謹依貴公司的來信，附上敝公司服務的價目表。
I have enclosed our catalogue, listing various types of translation services and related fees.	謹附上敝公司的目錄，內含多項翻譯服務以及費用說明。
As per your request, we are sending you a catalogue which contains product information of our company.	根據您的需求，我們已寄給您含有敝公司產品資訊的目錄了。
I have sent the latest catalogue to you this morning.	今天早上，我已經將我們最新的產品目錄寄給您了。

簡短介紹自家公司

I would like to give you a brief introduction.

我想要做個簡短的介紹。

We are extremely proud of our smartphones and the design of the new tablet PC.

我們對於敝公司出產的智慧型手機以及平板電腦的設計非常有信心。

The smartphones are selling wonderfully at some of your branches.

我們所生產的手機,在貴公司的某些分店銷售得極好。

Our service is recently utilized by many famous companies.

已有許多知名公司使用敝公司的服務。

We have translated our catalogue into different languages, and will send you a copy of the English version before the end of June.

敝公司的產品目錄已翻譯成多國語言,我們將會在六月底前寄英文版的目錄給您。

I am sure that our service will accelerate your profits.

我確信敝公司的服務將能增進貴公司的獲利。

關於目錄的補充

We already included your info in our mailing list, and you will be receiving our latest catalogue in no time in the future.

我們已將您的信箱加入我們的收件列表,新目錄一出來,您就會立即收到相關資料。

You can sign up on our website, and get our PDF catalogue by e-mail.

您可以登錄我們的網站,並接收 PDF 版的型錄。

A list of suppliers of our products can be found on page 3.

第三頁列出了所有供應商的名單。

If there is anything you can't find in our catalogue, please let us know.

如果有任何您需要的產品未出現在目錄中,請告知我們。

信件結語

We would appreciate your comments after reviewing our products .	看過我們的產品後，請不吝給予意見，謝謝您。
If you have any questions about our products, please let me know at any time.	如您有任何產品上的疑問，請隨時告知。
I am more than happy to answer any questions.	我很樂意回答任何疑問。
We will answer any questions you have about our products.	我們會回覆您對於產品的所有疑問。
Please contact me if you have any questions or would like some advice.	如果您有任何問題或需要一些建議，請與我聯絡。
Please contact us if you would like to order any of these items.	如果您有想要訂購的產品，請聯絡我們。
Your order will be proceeded immediately.	我們會立即處理您的訂單。
We will be happy to give you a special offer.	我們會給您特別優惠。
Please note that we offer a 15% discount on orders over US$2000 within one year.	在一年內，只要是超過兩千美元的單筆訂單，都將享有 15% 的折扣。
Once again, thank you for your interest in our products.	再一次感謝您對本公司產品的支持。
Thank you for your support.	謝謝您對敝公司的支持。
I will be in contact with you to see if you have any questions.	我會與您聯繫，看您是否有任何疑問。

Part
1
入門篇

Part
2
求職篇

Part
3
客戶往來篇

Part
4
人際互動篇

3-03 索取試用樣品

撰寫「索取試用樣品信」時,首先要表明身分及來信目的,接著有禮貌地詢問欲索取的樣品,強調提供樣品會對商務往來更有幫助;最後,若索取之樣品為試用後需要歸還者,請於信末註明將會歸還。

From:	Kevin Zheng
To:	Jessica Owen
Subject:	Request for Samples

Dear Jessica:

This is Kevin Zheng from Homey Corp. We are very interested in your goods, and would like to request for some samples of your new style jeans for women. Please let me know if possible.

We look forward to your prompt reply.

Yours faithfully,

Kevin Zheng

潔西卡您好:

我是家美公司的鄭凱文,我們對貴公司的產品非常有興趣,不知貴公司是否方便提供新款女性牛仔褲的樣品呢?若可以的話,麻煩您再跟我說。

期待您的迅速回覆。

鄭凱文 謹上

Many of my customers are interested in your new line of swimming suits, especially in the quality.	許多顧客對您最新款的泳裝非常有興趣，也想了解其品質。
We would be obliged receiving your catalogue with some samples.	如能收到您的目錄以及一些樣品，我們將感激不盡。
We would like to request a few samples before we place confirmed orders.	正式下訂單前，我們希望能索取一些樣品。
Before placing an order, may I request some samples of your goods?	在決定訂單前，我能先向您索取一些樣品嗎？
Before placing a confirmed order, I would be grateful if you could send us some samples of your new selection of children's shoes.	在下訂單之前，如果貴公司能夠提供最新系列的精選兒童鞋款，我們將非常感謝。
Could you please send us some free samples?	煩請寄送一些免費的樣品。
If available, please send us some samples.	如果方便的話，請寄給我們一些樣品。
We should find it most helpful if you could send us samples of your products.	如果貴公司願意提供產品的樣品，將會對敝公司有極大的幫助。
We would be very grateful if you could send us some samples.	若您可以寄來一些樣品，我們將非常感激。
Any of the items unsold would be returned.	未售出的產品將會歸還。

Part
1
入門篇

Part
2
求職篇

Part
3
客戶往來篇

Part
4
人際互動篇

3-04 提供試用樣品

於「提供試用樣品」的信件中，應告知提供的樣品數量，並強調商品之優點，表明期待能與對方合作。信末也應主動提供聯絡方式，以便對方詢問相關資訊。

From:	Jessica Owen
To:	Kevin Zheng
Subject:	Re: Request for Samples

Dear Kevin:

Thank you very much for your mail of May 20 about jeans for women. We are **glad**[1] to send you five pairs of samples you requested.

Our **fashionable**[2] style meets the **trend**[3], and we are certain that our **design**[4] **appeals**[5] to many customers.

Thank you again for your interest in our products. We look forward to having your **order**[6] soon, and please feel free to contact us if you have any questions.

Sincerely yours,

Jessica Owen

凱文您好：

感謝您於五月二十日來信詢問有關女性牛仔褲的詳情，我們很樂意寄送五件樣品給貴公司參考。

我們的服飾向來符合潮流，所以我們相信我方的設計能吸引許多買家。

再次感謝您對本公司的商品感興趣，我們期待不久後就能收到您的訂單。如有任何問題，歡迎隨時與我們聯繫。

潔西卡・歐文 敬上

1 glad 形 樂意的　　2 fashionable 形 流行的　　3 trend 名 趨勢

4 design 名 設計　　5 appeal 動 有吸引力　　6 order 名 訂購

複製 / 貼上萬用句 Copy & Paste

Thank you again for your interest in our products.	再次感謝您對本公司的商品感興趣。
We thank you for your inquiry of April 20.	謝謝您四月二十日的來信。
I was very pleased to receive your request of May 20.	很高興收到您五月二十日的請求。
We have sent a copy of our catalogue together with some samples of jeans.	我們已經寄出一份型錄，以及幾件牛仔褲的樣品。
We are forwarding some samples of the various women clothing as you requested.	根據您的要求，我們已寄出多款女性衣物的樣品。
We are providing you four samples of our most popular dresses.	我們提供四款洋裝的樣品給您，這四款是本公司最受歡迎的商品。
We are glad to send you samples of our goods you inquired.	我們很樂意將您要求的產品樣本寄給您。
I am pleased to send you a full range of samples of our new sneakers.	很高興能寄給您本公司最新款式球鞋的全系列樣品。
Please have a look of the samples.	煩請參考樣品。
I hope you will find our products suitable for you.	希望我方產品符合您的需求。

Part
1
入門篇

Part
2
求職篇

Part
3
客戶往來篇

Part
4
人際互動篇

3-05 詢問服務項目的報價

詢價時,可用直接但禮貌的態度書寫。信件開頭可先簡述自己是如何得知對方公司,並且向對方索取價目表、商品型錄、樣品等作為參考。

From:	David Lynch
To:	trans4you@service.net
Subject:	Translation Services

Dear Sir or Madam:

This is David Lynch from New Year Press Inc. We have seen your advertisement in your catalogue No.81, and would be grateful if you could send us details about your translation services.

Thank you for your prompt response to this enquiry.

Yours sincerely,

David Lynch

敬啟者:

我是新年出版的大衛・林區。我們在貴公司第八十一期的型錄上看到廣告,若您願意提供詳細的翻譯服務資訊,我們會十分感激。

非常感謝您能即時回覆詢問。

大衛・林區 謹上

複製 / 貼上萬用句 Copy & Paste

在何處得知對方資訊

 We have seen your advertisement on the newspaper.

我們在報紙上看到您刊登的廣告。

We would be thankful if you could send us the estimated costs.	若您能提供預估的價格，我們將感激不盡。
We would like to request more information about your service.	我們希望能了解更多貴公司的服務內容。
You were referred by one of our clients.	我們的一位客戶向我們推薦貴公司。
Your company was recommended by Leader Consulting Company.	貴公司是由領導諮詢公司所推薦。

需要的服務類型

We are interested in placing advertisement on the newspaper.	我們想要在報紙上刊登廣告。
We are interested in placing advertisements on the buses.	我們想要在公車上設置廣告。
We would like to buy some advertising place on different business websites.	我們想要在各大企業的網站上購買廣告位置。
We intend to purchase advertisement in some business magazines.	我們想要在商務雜誌上刊登廣告。
We are very interested in your service for cleaning the window in our office building.	貴公司所提供的服務當中，我們想了解清洗大樓櫥窗的業務。
We are recently developing our website and are planning to translate the website into four languages apart from Chinese.	近期內，本公司會將正在更新的網頁翻譯成中文以外的四種語言。
Is your company with the capability of translating Japanese into English?	不知道貴公司是否有提供日翻英的服務？

請對方提供資訊

Please send us your advertising rate card.

請回覆告知您的廣告報價。

Please write or call me to provide further information.

煩請寫信回覆，或是以電話與我聯繫，以提供更多資訊。

I would be grateful if you could send us your brochure and price list about your translation services.

如果您能寄送貴公司翻譯服務的內容以及價目表，我會非常感謝。

Would you be kind to send us an estimate for the above translation project?

可以請您寄給我們上述翻譯服務的估價嗎？

We would be pleased if you could send us shipping estimate details.

希望您能寄詳細的運費報價給我們。

信件結語

Please feel free to call us if you need more specific information about our inquiry.

如需詳細資訊，歡迎隨時來電。

Please let us know if we can provide further information.

如果需要更多資訊，請讓我們知道。

It is a pleasure for us to serve you.

為您服務是我們的榮幸。

I look forward to hearing from you.

期待您的消息。

We are eagerly waiting for your reply.

我們熱切等待您的回覆。

3-06 回覆服務項目的報價

回覆報價內容時，要再次強調己方公司的優勢，以及可以提供給對方的服務內容，並於信中感謝對方選擇自家公司的服務。

From:	Wendy Scott
To:	David Lynch
Subject:	Re: Translation Services & Fees

Dear David:

Thank you for your mail of Sep. 22 enquiring about our translation services.

BILINGUA Translation Services offers a full range of translation services to help you develop both website content and sales requirement. I have the pleasure to enclose our latest brochures and price list. You will find our prices highly competitive.

I look forward to contacting you soon.

Yours sincerely,

Wendy Scott

大衛您好：

謝謝您九月二十二日來信詢問本公司的翻譯服務。

「雙語翻譯服務公司」提供多樣化的翻譯服務，幫助您建立網頁內容以及拓展業務。隨信附上最新的目錄與價目表給您，敬請參考本公司的價目，您會發現我們提供的價格非常具有競爭力。

期待能於近日與您聯繫。

溫蒂・史考特 敬上

信件開頭：回覆對方的來信

🖱 We appreciate the opportunity to bid the office window cleaning case.

關於清洗大樓櫥窗的招標，我們感謝您提供我們投標的機會。

🖱 Thank you for choosing our company to place advertisement for you in the magazine.

感謝您選擇本公司替貴公司刊登雜誌廣告。

🖱 We are pleased that you choose our company to print catalogues for you.

感謝您選擇本公司替您印刷目錄。

請對方提供進一步的資訊

🖱 We would like to ask you three questions as follows:

我們想請您回覆下列三點問題：

🖱 We will need more detailed information before making a confirmed quotation.

在此需要您提供更詳細的資訊，以便給予您精確的報價。

🖱 Your answers will help us bid accurately.

您的答覆可以幫助我們做精確的投標。

信件結語

🖱 We offer a free consultation for first-time clients.

我們提供新客戶一次免費的諮詢服務。

🖱 We appreciate to work with you.

我們很感激能有與您合作的機會。

3-07 詢問產品報價

　　詢價信基本上會包括以下內容：一、向廠商說明從何處得知廠商的資料；二、自我介紹並說明與其聯繫的理由，例如「我們看到貴公司的產品廣告」，或「我們收到客戶對貴公司產品的詢問」等。清楚表明詢問項目與貨品的需求量，以便對方報價；三、要求廠商報價和說明付款條件，並要求提供產品的其他資訊；四、說明若價格和品質符合需求，將會下單；五、最後註明會靜候佳音，表示期待對方的回信。

From: Anna Jackson

To: Ms. Lee

Subject: Order 20 New Printers

Dear Ms. Lee:

We are interested in ordering 20 new printers for our new office in Taichung. Could you please send us a quotation and provide the details of specification sheet? Your prompt reply will be appreciated.

Yours sincerely,

Anna Jackson

李女士您好：

我們剛在台中成立新公司，因此想向貴公司訂購二十台印表機。可否請您寄估價單給我們，並附上詳細的商品規格資訊？感謝您的即時回覆。

安娜·傑克森 敬上

複製 / 貼上萬用句 Copy & Paste

點出聯繫目的

We have seen your advertisement in Washington News and are interested in your printers.	我們在《華盛頓日報》上看到貴公司的廣告，並對您的印表機有興趣。
We saw your men's suits at the Taipei Fashion Show.	我們在台北時裝展看到貴公司的男性服飾。
We have received an inquiry for your products from one of our trade connections.	我們接到顧客詢問貴公司的產品。
We received many inquiries for the following goods:	我們接到許多關於下列產品的詢問：
I would like to make an inquiry about your goods.	我想詢問貴公司的產品。
Thank you for the catalogue that we received last Tuesday.	謝謝您上週二寄達的型錄。

請對方報價

We are interested in purchasing your computers and would like to know some details.	我們有興趣購買您的電腦產品，因此想了解更多細節。
Would you please send us your quotations for summer clothing?	煩請您將夏季服飾的報價單寄到敝公司好嗎？
We would be grateful if you could send us your recent prices.	如果您能寄送最新的價目表，我們會很感激。
We are interested in your products and would like to know the following:	我們對您的產品有興趣，並想了解以下資訊：
We would like to know the current prices for your products.	我們想要了解貴公司目前提供的產品價格。

We would like to know the price of your e-Books.	我們想要了解貴公司電子書的價位。
Please quote for the latest prices for your new style ovens.	請您提供新型烤箱的最新報價。
We would be pleased if you could send us an estimate for your digital cameras.	如果您能寄送貴公司數位相機的報價單，我們會很感激。
We are writing to request some information about your laser printers.	我們想要進一步了解貴公司的雷射印表機。
We are interested in importing Swiss chocolate and would like to receive the latest price list.	我們對進口瑞士巧克力有興趣，希望能收到最新的價目表。
We are interested in importing Italy wine and would like to receive a copy of your export price list and export terms.	我們對進口義大利酒有興趣，希望貴公司能提供出口價目表與出口條件給我們。
Please send us full details of your products and state your earliest delivery date, terms of payment, and discounts.	請一併告知貴公司產品的詳情、最快交貨日、付款條件以及優惠。

開始談折扣

The fashion design you presented for the young men would be suitable for the market.	貴公司所展示的男性時尚設計非常具有市場性。
We may place an order with you if your prices are competitive.	如價格有競爭力，我們也許就會向貴公司訂購。
We would like to confirm your best business terms regarding discount.	我們想進一步確認貴公司最佳的折扣條件。
Would you please let us know on what term you will give us some discount?	煩請告知折扣的相關條件好嗎？

We would like to know your best business terms and discount.	請將您最佳的折扣條件告訴我們。
Please send us your best quotations for these products.	請告知產品最優惠的價格。
We would be highly obliged if you send us a copy of your catalogue with your best terms and lowest prices.	如果您能寄送型錄以及最優惠價格之條件，我們會非常感激。
We would ask you to quote at a competitive rate.	請您儘量壓低報價。
If we place a large order, will you give us some discount?	如訂單量大，貴公司可否給予折扣優惠？
How much discount can be offered if we place a large order?	如訂單量大，貴公司可以給予多少折扣呢？
If there are any special offers or opportunities to obtain the above items at lower prices, please inform us.	如果上述產品有機會降價，煩請告知。

除了報價以外的資訊

Please also let us know your shipping and insurance costs.	也請告知運費以及保險的價格。
Would you send us some samples with quotation?	能否請您寄送報價單的同時，附上一些樣品給我們呢？
Please let us know your minimum order quantity and conditions.	請您告知訂購時的最低要求數量以及相關條件。
We would be appreciated if you would send us details regarding prices, terms of payment, delivery schedules and possible discount.	若能寄來有關價位、付款條件、貨運日程及相關折扣等詳細資訊，將不勝感激。

3-08 回覆產品報價（簡略版）

此處的範例為非正式報價，因此只在郵件中附上產品型錄以及價目表。正式的報價信請參閱「Unit 4 報價篇」。（若報價涉及「如有價格變動不另行通知」的內容時，記得要在信中提前告知客戶。）

From:	Nancy Lee
To:	Anna Jackson
Subject:	Re: Order 20 New Printers

Dear Anna:

Thank you for your inquiry about our new printers. We are pleased to attach our latest catalogue and the price list. Please let us know if there are any other inquires. Thank you!

Yours sincerely,

Nancy Lee

安娜小姐您好：

感謝您對本公司印表機的詢價，附件為最新的產品目錄以及價目表。若有其他問題，請再讓我們知道，謝謝！

李南西 謹上

複製 / 貼上萬用句 Copy & Paste

信件開頭

🖱 **Thank you for your enquiry on May 20.**	謝謝您於五月二十日的來信詢問。
🖱 **Thank you for your enquiry of April 1 for further information of our machines.**	謝謝您於四月一日的來信，詢問本公司機械的詳細資訊。

Thank you for choosing our company for your office equipment supply.

感謝貴公司的青睞，選擇我們為辦公室設備的供應商。

We appreciate your interest in our products.

謝謝您對我們的產品感興趣。

May we ask you how did you know about our company?

請問您是如何得知敝公司的呢？

回覆產品報價

Enclosed is a copy of the price list and estimated costs.

附件為價目表與估價單。

Enclosed please find our price quotation.

附件為我們的報價單。

With reference to your enquiry, we are pleased to enclose our price list and terms of payment for your consideration.

為回覆您的詢問，我們附上價目表和付款方式，供您參考。

In compliance with your request, we are sending you our quotation for bicycle No. 55.

根據您的請求，我們已寄 NO. 55 型號的腳踏車報價給您了。

Please see the attached file for photos and price quotation for our products.

請參照附件的產品圖片以及估價表。

I am also including a specification sheet about the laser printer for your technical questions.

同時附上雷射印表機的規格表，希望能夠解答您的問題。

We are pleased to quote as follows:

我們的報價如下：

The prices quoted are valid until December 31.

此報價只提供至十二月三十一日。

The price includes delivery fee.

此報價已含運費。

The price does not include delivery and insurance costs.

此報價不包含運費以及保險費用。

提供產品資訊 & 樣品

We are greatly glad to provide you with the information you requested.	我們很開心能提供您所需的資訊。
We are more than happy to supply you with women T-shirt.	我們很開心提供您女性T恤。
As to the samples, we have sent them to you separately by airmail.	我們已另外郵寄樣本給您。

可給予的折扣優惠

We give a discount of 10% on orders of over 100 pieces of printers.	若訂購一百台以上的印表機，我們會提供百分之十的折扣。
As you are a valued customer, we will reduce prices by 5%.	由於您是我們的重要客戶，我們將給予九五折的優惠。
This offer will be withdrawn if not accepted within five days.	特別優惠將保留五天。
A 10% discount is offered on payment within two weeks.	兩週內付款有九折優惠。
We are prepared to offer special terms for orders over US$1,000 by the end of this month.	本月底前消費超過一千元美金的顧客，我們會提供特別的優惠。
To meet your demand, we will offer a special discount of 10% off for any order you place within 7 days start from today.	為滿足您的需求，即日起七日內，我們提供九折的優惠價格。
In reply to your enquiry of A700 e-Book, we have pleasure in making the following offer to you.	關於您所詢問的A700號電子書，我們提供以下的特別優惠給您。
The discount is valid for one week, from now until December 15, 2019.	優惠期限為一週，至二○一九年的十二月十五日截止。

補充報價的相關細節

🖱 **The offer is subject to our final confirmation.**

報價以本公司的最後確認為主。

🖱 **All prices are subject to change without notice.**

所有的價格更動不另行通知。

信件結語

🖱 **We hope you are satisfied with our estimate.**

希望您滿意我方的報價。

🖱 **It is our honor to be your office equipment supplier.**

成為您的辦公室設備供應商是我們的榮幸。

🖱 **We are confident that our goods are both excellent in quality and reasonable in price.**

對於敝公司產品的品質及合理的價位，我們深感自信。

🖱 **I am confident that you will find our digital cameras attractive and functional.**

我有自信，我們的數位相機不僅吸引人，也很實用。

🖱 **As the stocks are low, we hope you would send us your orders as early as possible.**

由於存貨量不多，希望您能儘早下訂單。

🖱 **If you need any further information about our products, please do not hesitate to let us know.**

如果您需要任何其他產品的相關資訊，請不吝告知我方。

🖱 **We are confident that our goods meet your requirements.**

我們有信心，我方的產品能符合您的需求。

🖱 **We assure you that your order will receive our careful attention.**

我們可以保證，您的訂單會得到我們的細心關切。

3-09 詢問產品現貨

某些產品的銷售有季節性，可在詢問現貨時告知需要出貨的日期與商品的需求量，並可額外說明最慢可接受的到貨時間，以及可否接受對方先出一部分的貨。

From:	Greg Lee
To:	Ms. Tsai
Subject:	Order of Luxury Bed Sheets

Dear Ms. Tsai:

We would be grateful if you would send us patterns and prices for your Luxury bed sheets. Please also inform us whether you could supply these goods from stock, because we need them before the Christmas season starts.

Yours sincerely,

Greg Lee

蔡小姐您好：

若您能提供我們豪華床罩的樣式和價格，我們將十分感激。也請告知是否有存貨可供應，因為我們在聖誕季開始之前就需要拿到這批商品。

葛瑞格 · 李 謹上

 **複製 / 貼上萬用句 Copy & Paste**

詢問庫存量 & 相關

 I am writing to request about your stock availability.

我想了解貴公司的庫存量。

Please inform us how many quantities your company can supply from stock.	請告知貴公司能提供的存貨數量。	Part 1 入門篇
Please check your inventory to see if you have 1,000 bed sheets for delivery.	請確認您的庫存是否能提供一千件床罩。	Part 2 求職篇
If possible, we would like to place an order for 30 laser printers.	如果貴公司有足夠的庫存，我們想要訂購三十台雷射印表機。	
If possible, we need 10 more for another delivery.	若庫存量容許的話，我們想要再加購十台。	Part 3 客戶往來篇
If we place a large order, would you provide us enough goods?	如果訂單量很大，貴公司是否有足夠的庫存量能提供？	
What type of model do you have in your stock?	貴公司的存貨尚有何種型號？	Part 4 人際互動篇
Will you please send us the date of delivery and quantities you can supply from stock?	請告知您能出貨的日期及現貨能供應的數量。	
Please check your inventory to see if you have 100 digital cameras for delivery by July 20.	請查詢貴公司庫存，以確定是否能在七月二十日前寄出一百台數位相機。	
Please let us know your earliest date of delivery.	請告知您最快能什麼時候出貨。	
Please also let us know if your company is prepared to grant a ten-percent discount.	請告知貴公司能否給予百分之十的折扣。	

信件結語

If you can supply from stock, we will give you an order by return.	若貴公司供貨，我們便會向您訂購。
Upon receipt of your information, we will place an order with you soon.	一接到您的回覆，我們便會儘速向您訂貨。

3-10 回覆產品現貨

在「回覆產品現貨」的信件中，要明確告知產品是否有現貨供應，以及貨運的時間，並於信中告知客戶提早下訂單便可盡快處理出貨，好讓對方更快下單。

From:	Ella Tsai
To:	Greg Lee
Subject:	Re: Order of Luxury Bed Sheets

Dear Greg:

Thank you for your e-mail. We would like to inform you that we have 2,000 **bed sheets**[1] **in stock**[2]. Those **items**[3] in stock are **inclusive**[4] of **tax**[5]. We would be happy to have our stock **status**[6] e-mailed to you from now on. We deliver three times a week. Therefore, since we have your order, we would deliver the items as early as possible.

Yours sincerely,

Ella Tsai

親愛的葛瑞格：

感謝您的來信。我們想要通知您庫存尚有兩千件床罩，這批現貨的價格都已含稅。今後本公司會提供商品庫存量給您參考。我們每週出貨三次，因為我們已經收到您的訂單，所以會盡快替您出貨。

蔡艾拉 敬上

key words ▲ E-mail 關鍵字一眼就通

1 bed sheet 片 床單　　**2** in stock 片 有現貨　　**3** item 名 物品

4 inclusive 形 包含的　　**5** tax 名 稅金　　**6** status 名 情況

Part
1
入門篇

Part
2
求職篇

Part
3
客戶往來篇

Part
4
人際互動篇

複製 / 貼上萬用句 Copy & Paste

庫存量充足

- We are now well-stocked with those products you requested.

  您所需要的產品目前備貨充足。

- We are ready to deliver any quantity of table lights from stock.

  我們隨時可以供應桌燈。

- All models can be supplied from stock.

  所有款式皆有存貨。

庫存量有限

- We have checked our supply of the model you requested.

  我們已確認您所需型號的庫存量。

- We have only a limited quantity of goods now.

  該商品目前的庫存量有限。

- Our stock level is getting low; therefore, we would ask you to place your order as soon as possible.

  由於庫存量不多，我們希望您能儘速下訂單。

無存貨 & 產品絕版

- The digital camera is now out of stock.

  這款數位相機已無存貨。

- We no longer provide children's shoes.

  我們已無販售童鞋。

提供優惠資訊

- As we are going to clear our stock, we are selling our goods at a 15% discount.

  我們即將出清存貨，商品將以八五折的優惠價格出售。

- As we are closing down one of our shops, you could subtract 20% from the prices in the price list.

  本公司即將關閉一家分店，因此，商品定價將打八折。

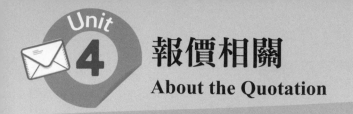

報價相關
About the Quotation

4-01 正式報價

　　報價信通常分為三段。第一段感謝對方來信,並簡單複述對方的詢價;第二段提出報價並說明優惠;第三段提醒對方儘速下訂單,以免錯失優惠。報價時,須清楚寫明商品名稱、數量、價格、折扣、運送時間、付款方式以及報價有效期限等訊息。

From:	Kelly Lai
To:	Mr. Wang
Subject:	Quotation

Dear Mr. Wang:

Thank you for your **inquiry**[1] of May 30, **herewith**[2] we quote our best prices as shown **below**[3]:

100 *Item No.201 ....US$20

100 *Item No.204 ....US$22

100 *Item No.208 ....US$26

Minimum[4]: 100 sets

We can give a 5% discount on all orders of $500 and more, and the prices are **valid**[5] until August 31st, 2018.

We look forward to your **comments**[6] or your first order confirmation soon.

Sincerely yours,

Kelly Lai

王先生您好：

謝謝您五月三十日來信詢價，以下為我們的報價：

品項 201*100 份	20 美元
品項 204*100 份	22 美元
品項 208*100 份	26 美元

最小訂購量為一百份。

訂購價格超過五百美元時，我們將給予百分之五的折扣；報價有效期限為二〇一八年八月三十一日。

希望不久後就能收到您的建議或首次訂購。

賴凱莉 敬上

key words ▲ E-mail 關鍵字一眼就通

1 inquiry 名 詢問
2 herewith 副 隨函
3 below 副 在下面
4 minimum 名 最小量
5 valid 形 有效的
6 comment 名 意見

From:	Emma Evans
To:	Albert Green
Subject:	Quotation

Dear Mr. Green:

Thank you for your e-mail of February 15, expressing your interest in our goods. We are pleased to offer you our best quotation as follows:

100 *Item No.202 ......US$20 ....US$2,000

100 *Item No.205 ......US$22 ....US$2,200

100 *Item No.207 ......US$26 ....US$2,600

$6,800

5% Discount	$340
Net price	$6,460
Freight (Sydney to Paris)	$200
Insurance	$120
Total	$6,780

Shipment: Within 2 weeks of receiving order.

Valid for 20 days

We hope to receive your order soon.

Sincerely yours,

Emma Evans

格林先生您好：

謝謝您五月十五日來信表示對我們的產品有興趣，以下為我們的最佳報價：

品項 202*100 份	2,000 美元
品項 205*100 份	2,200 美元
品項 207*100 份	2,600 美元
	6,800 美元
折扣 5%	340 美元
淨價	6,460 美元
運費（雪梨到巴黎）	200 美元
保險費	120 美元
總計	6,780 美元

貨運：收到訂單後的兩週內送達。報價有效期為二十天。

盼能早日接到您的訂單。

艾瑪・伊凡斯 敬上

複製 / 貼上萬用句 Copy & Paste

回覆詢價信件

It is our pleasure to receive your inquiry.	能夠接到您的詢問是我們的榮幸。
We are pleased to receive your inquiry.	我們很開心能收到您的詢問。
The prices you requested are as follows:	您所需要的報價如下：
We are pleased to quote as follows:	我們很開心能提供以下報價：
Please kindly refer to our quotation as following:	請參考以下報價：
Attached please find our best quotation for your reference.	附件為我們最優惠的報價。
The quotations and packing details of the items that we discussed are as below:	我們討論過的產品報價以及包裝細項如下：
Please refer to our price list and you will find our best prices.	請查閱我們提供的最優惠報價單。
In reply to your letter of mail of April 28, it is our pleasure to send you a quotation for our products you required.	回覆您四月二十八日的來信，我們很榮幸附上您所需要的產品報價。
We are glad to submit our lowest prices for the products you inquired.	我們很榮幸附上您所需產品的最低報價。
We are pleased to provide you with the enclosed price list for our products.	很高興能提供您我方的產品報價。

關於報價之補充

The prices quoted in the attached price list are on CIF terms, including postage and insurance.
附件之報價單為包括運費、保險費在內之到岸價格。

Shipment and insurance costs are to be paid by the buyer.
運費以及保險費用將由買方負擔。

The prices include packing and delivery fees.
此報價包含包裝以及運抵貴公司的費用。

We may raise our prices because of the increased cost of raw materials.
由於原物料價格上漲，因此，我方的產品價格也有可能跟著調漲。

As our supplier has raised his price, our prices will therefore go up next week.
本公司的供應商已調漲價格，因此，我們下週也會隨之調漲。

The prices quoted here are subject to change.
報價有可能會調整。

The quotation does not represent a final order.
此報價並非最終報價。

Prices are subject to change without notice.
價格變動不另行告知。

The offer is subject to the final quote from our shipping partner.
最終報價將會根據貨運廠商的費用而定。

This quotation is based on the currency exchange rate: USD 1.00 = NTD 30.31.
報價根據匯率：一美元兌新台幣三十點三一元。

The quotation is based on the exchange rates on June 20.
此報價是根據六月二十日的匯率而定的。

Please note that we have quoted our most favorable price and are unable to entertain any counteroffer.
此報價已為最優惠的價格，恕難還價。

期間內的特定報價

🖱 This offer will expire on July 31.

此報價七月三十一日前有效。

🖱 The payment needs to be made within 30 days.

須於三十日內付款。

🖱 The prices quoted are valid for 14 days and include shipping costs.

報價含運費，有效期限為十四天。

提供優惠方案

🖱 We can offer you a price of $23 per item.

每項產品的報價為二十三元。

🖱 If your order quantity is more than 1,000 pcs, we can offer you a 25% discount of the list price.

如果您的訂購量超過一千件，我們可以提供您七五折優惠。

🖱 We offer a 5% cash discount.

我們提供百分之五的現金折扣。

🖱 We offer a 5% discount on orders over $500.

訂購量超過五百元者，我們將提供九五折的優惠。

🖱 Orders over $2,000 are subject to negotiation.

訂購量超過兩千元則可議價。

🖱 We would allow you a discount of 5% if you place your order before July 20.

如您能於七月二十日前下訂單，我們願意給您九五折的優惠。

🖱 We can grant you a 15% discount on repeat orders.

後續訂單可享八五折優惠。

🖱 Our price will decrease as we need to turn over the remainder of the stock.

由於我們準備出清存貨，因此，報價將會隨之調降。

供貨情況

- Most models are obtainable from stock except the blue ones, of which the large sizes have been sold out.

  大尺寸的藍色款已售罄,其餘產品皆可出貨。

- We have the goods in stock and will ship them immediately upon receiving your order.

  我們的存貨充足,一接到您的訂單就能馬上出貨。

- We can grant delivery within one week after receiving your order.

  在收到訂單後,我們保證會於一週內出貨。

- If you can make your payment within 10 working days, we would deliver your goods soon.

  如果您可於十個工作天內付款,我們會替您快速出貨。

- If you intend to place a large order, please call our sales manager at 0800-112-233.

  若您的訂購量龐大,請來電聯繫我們的業務經理,電話請撥:0800-112-233。

- Please place your order as soon as possible.

  請儘早下訂單。

- Any orders you place with us will be arranged immediately.

  一收到您的訂單,我們就會立即處理。

- Any orders of yours will receive our best attention and prompt service.

  我們會留意您的訂單,並盡快處理。

- Orders will be packed and shipped two weeks after orders received.

  貨品將於下訂後兩週內包裝並寄出。

- We need to know if you need the products immediately, because there are other customers who are interested in them as well.

  我們必須確認您是否立即需要此項產品,因為其他客戶也對此產品有興趣。

說明支付方式

We accept bank transfers, check, credit card or money orders.	我們接受銀行轉帳、支票、信用卡或匯票。
For first-time buyers, we require payment in advance.	我們希望首次購買者能預付貨款。
We won't accept payment in cash on delivery.	我方不接受貨到付款的支付方式。
We insist on payment in cash on delivery without allowing any discount.	我方堅持貨到付款，不打任何折扣。

重申合作意願

We are confident that our offer is competitive compared to other manufacturers.	和其他廠商相比，相信我方的報價很有競爭力。
Hope that our offer can meet your demand, so that you may confirm your order soon.	希望我方的報價能讓您滿意，並儘速確認訂單。
We hope you find our offer satisfactory, and can confirm your order soon.	我們希望您對報價滿意，並儘早提出訂單。
We look forward to your initial order.	我們期待您的初次訂購。
We look forward to receiving a trial order from you.	我們期待接到貴公司的試用訂單。
We are pleased to cooperate with you soon.	我們很開心即將能夠與貴公司合作。

4-02 針對報價進行議價

要進行議價時，可於信中表示現今市場環境競爭激烈而請求對方降價；也可提出若本次訂購順利，後續將有長期配合的可能，如此較有說服力。

From: John Wang

To: Kelly Chen

Subject: Re: Quotation

Dear Kelly:

Thank you for your quotation of **January**[1] 12, 2018.

According to our **initial**[2] research, the **leather**[3] shoes are very popular in Europe; **hence**[4] we'd like to place some orders. However, we need the most competitive price. If you would lower $5 per item, we would like to **place an order**[5] with you for 2,000 pairs of leather shoes. All the other **terms**[6] are acceptable.

I hope to receive your confirmation soon.

Best regards,

John Wang

凱莉您好：

謝謝您於二〇一八年一月十二日的報價。

根據我們的初步調查，皮鞋在歐洲非常受歡迎。但是，我們仍然需要具競爭力的價格。若貴公司每件商品能降價五元的話，我們願意訂購兩千雙皮鞋。除了價格之外的其他條件，我們都能接受。

盼能儘早接到您的確認。

王約翰 敬上

Part
1
入門篇

Part
2
求職篇

Part
3
客戶往來篇

Part
4
人際互動篇

key words ▲ E-mail 關鍵字一眼就通

1 January 名 一月　　2 initial 形 初步的　　3 leather 形 皮革製的

4 hence 副 因此　　5 place an order 片 下訂單　　6 terms 名 條款（常複數）

🖅 複製 / 貼上萬用句　Copy & Paste

表示已審核過報價

We have discussed your offer carefully.	我們已仔細討論過您的報價。
We have carefully considered the quotation in your mail.	我們仔細討論過您來信中的報價。
Your quotation has been reviewed.	我們已檢閱您的報價。
We have studied your offer in detail.	我方已詳細檢閱您的報價。
We have received quotations from other companies.	我方也從其他公司取得報價。

告知報價過高

We also got some lower offers from other companies.	其他公司提供了較低的報價。
Because of the current exchange rate, many other manufactures would offer us lower prices.	由於目前匯率的變動，許多製造商願意提供我方較低的報價。
Your products are more expensive than those in the local's.	貴公司的產品比起當地產品要貴得多。
We understand your goods are of excellent quality.	我們了解貴公司的產品具備優良的品質。
We don't think the price is competitive at this time.	我們覺得此價格不具競爭力。

Your prices are the highest ones.	您的價格為最高的報價。
Your prices are higher than other dealers.	您的價格高於其他業者。
We find the price higher than our expectation.	您的價格已經超出我們的預期。
We think your quotation is unreasonable.	我們覺得您的價格不合理。
Our customers are not satisfied with the price.	我們的客戶並不滿意這個價格。

表達難處 & 態度

The value of Euro is increasing.	歐元的價格上漲。
Our profit would be very little.	我們的利潤很少。
These prices are not competitive enough in the market here.	目前的報價在這個市場上沒有足夠的競爭力。
We have to compete with other manufacturers.	我們必須與其他的製造商競爭。
The prices you offered would be difficult for us to promote the products in the market.	您提供的報價使得我方難以在市場上促銷產品。
In order to remain competitive, we need to reduce our prices.	為了保有競爭性，我們必須壓低產品的價格。
Please understand the high prices will turn our customers to other suppliers.	商品價格太高會導致我方的顧客轉向其他供應商，這一點要請您體諒。
We won't accept the prices you offered.	我們無法接受您的報價。
We would like to ask you to reconsider the price.	我們希望您重新考慮報價。

We are not able to confirm the contract because the prices are too high.	由於價格過於昂貴，所以我們無法下訂單簽約。
If your prices remain the same, we may need to purchase the products from China.	如果您的報價維持不變，我們可能就必須向中國訂貨。
If you cannot give us better prices, we would give up.	如果您無法提供優惠價格，我方將會放棄訂購。

要求對方降價

We hope you can allow us a 15% discount of the list price.	我們希望您能降價百分之十五。
We would like to ask you to reduce the quoted prices by 5%.	我們希望您能降價百分之五。
Would it be possible to lower your price?	不知貴公司是否能降低價格？
Is there any discount?	不知是否有折扣？
We hope you could give us a 15% discount from the price list.	希望您能提供百分之十五的折扣。
If the price per item is reduced by $5, we will place an order soon.	如果每項產品可以降價五元，我們會儘速下訂單。
If we place a large order, we hope that you would consider reducing the price.	如果我們下大量訂單，希望貴公司能考慮降價。
If we place orders on a regular basis, would you please grant us a 20% discount from the listed price?	如果我方定期下單，您是否可以提供百分之二十的折扣呢？
We would increase the volume of our order if you increase your discount from 5% to 10%.	如果貴公司將百分之五的折扣提升至百分之十，我們會增加訂單量。

有條件地接受報價

We will accept your offer if you can ship the goods within a week after receiving the order.

如果您可以在接到訂單後的一週內出貨，我們願意接受這個報價。

We would accept your offer if we are allowed to pay in installments.

如果可以分期付款，我們願意接受您的報價。

We will accept the price you offered if it includes delivery and insurance costs.

如果您的報價含運費和保險費用，我們將會接受這個報價。

We will place our order as soon as we receive your reply regarding the discount.

確認最終的折扣價後，我們將會儘速下訂單。

信件結語：表達感謝

Please reconsider it.

請您重新考慮。

Your kind help to provide more discount will be greatly appreciated.

您若能體諒並提供更多的優惠，我們會非常感激。

We hope you will accept our suggestions.

希望您能接受我們的建議。

Your kind support is highly appreciated.

非常感謝您的幫忙。

Thank you for your consideration.

謝謝您的考慮。

We look forward to hearing from you soon!

我們期待能早日得到您的回音！

Part
1
入門篇

Part
2
求職篇

Part
3
客戶往來篇

Part
4
人際互動篇

4-03 回覆議價

回覆潛在買家之議價時，可強調產品本身的優越性，或是由於原料以及匯差上的因素而無法接受殺價；除此之外，在長遠合作關係的考量下，亦可給予有條件的議價空間，如大量訂購則可提供折扣等。

From:	Vera Collins
To:	Mr. Batson
Subject:	Re-quote Products

Dear Mr. Batson:

We have read your e-mail with attention. We are very sorry that you didn't find our offer attractive. Although our products cost higher than similar goods, our quality is far better.

However, considering your situation, we will grant you a special discount of 5% on a first order for $2,000.

We look forward to your reply.

Best regards,

Vera Collins

貝特森先生您好：

我們已收到您的回信，非常遺憾您不滿意報價。雖然我們產品的價格較其他相似的產品高，但品質非常優良。

然而，考量到您的情況，若您的首批訂單達兩千元，我們願意給予百分之五的折扣優惠。

期待您的回覆。

薇拉‧科林斯 敬上

複製 / 貼上萬用句 **Copy & Paste**

回覆議價信函

🖱 Thank you for your reply.
感謝您的回覆。

🖱 Thank you for your mail regarding your counteroffer.
謝謝您來信議價。

🖱 Your counteroffer has been received with the best attention.
您的議價來信已受到關注。

🖱 We have received your mail regarding price reduction.
我們已收到您要求降價的回信。

接受對方的議價

🖱 We have decided to accept your request regarding price reduction.
我方決定接受您的降價要求。

🖱 We accept the terms you specified for this order.
我們接受您的議價。

🖱 Regarding your request for a 3% reduction in price, we will agree on that so to help you sell.
我們同意降價百分之三以幫助您銷售本產品。

🖱 We would like to reduce the price by $2 per item.
我們願意針對每項產品調降兩元。

🖱 In order to meet your request, we would offer you a special discount of 10%.
為了符合您的要求，我們願意提供您百分之十的折扣。

🖱 We have considered our future business relationship; therefore, we agree with your request.
為了未來的合作關係，我們同意您的要求。

表示已給予優惠價格

We are sorry to hear that you are not satisfied with the quotation.	我們對於您不滿意報價深感遺憾。
I am sorry that the terms we offered did not meet your requirements.	很遺憾我們的報價無法符合您的需求。
We are sorry to hear that you find our prices too high.	我們很遺憾您覺得我們的報價太高。
We have to refuse to lower the prices as you requested.	我們必須拒絕您的降價要求。
Unfortunately, we cannot give you the discount as you requested.	很遺憾，我們無法提供您要求的折扣。
Unfortunately, we cannot accept the price you requested.	很遺憾，我方無法接受您所建議的價格。
We can understand your situation.	我們了解您的情況。
We have quoted you with our best prices.	我們提供給您的，已經是最優惠的價格。
The price we offer in our last mail is the best price we can give.	我們已在上封郵件中提供最優惠的價格。
We have done our best to make our prices as low as possible.	我方已盡力壓低價格。
We hope you could understand that the discount is the maximum we can offer.	希望您能了解我們已經提供您最大的折扣。
We have already reduced the prices down by 5%.	我們已經價降百分之五。
The prices have been cut.	價格已經調降了。
We are sure the terms of our offer are very competitive.	我們確信這個報價很有競爭力。

We are afraid that we cannot make a better offer to comply with your request.	我方恐怕無法再根據您的要求調降價格。

表達難處 & 提升說服力

We hope to be able to meet your needs, but your counteroffer is too low for a small amount of order.	我們也希望能夠滿足您的需求，但就訂購量來說，您的議價過低。
NT dollar has appreciated. / NT dollar has depreciated.	台幣升值了。/ 台幣貶值了。
The prices of raw materials have increased.	原物料價格上漲了。
The cost of the goods is increasing for the rocketing cost of raw materials.	由於原物料價格上漲，因此產品的成本也隨之增加。
Considering the high production cost, we do not think our price is too high.	由於製造成本昂貴，我方不認為價格過高。
Our margin of profit is getting smaller.	我方的利潤空間越來越小。
Our profit margin won't allow us any concession.	因為利潤太少，我方無法讓步。
We have difficulties making any further reductions.	進一步降價對我方來說有困難。
The quality of our products is far superior to other similar ones.	我們產品的品質遠優於其他類似產品。
We are confident you will be satisfied not only with the quality of the goods, but also with our after-sales service.	我們相信您會對於本公司產品的品質以及售後服務感到滿意。

針對議價空間談判

I would like to offer a few alternatives.	我有其他的替代方案。

If you increase your order volume, we would be happy to allow you the 10% discount.	如果您增加訂購量，我們願意提供九折優惠。	
We hope you to increase your order to 2,000 pieces, which is our minimum production quantity.	我們希望您能增加訂購量至最小生產額兩千份。	
We can only offer 3% price reduction instead of 6% you requested.	我們只能提供百分之三的折扣，而非您要求的百分之六。	

信件結語：期待對方回覆

We hope you can reconsider our offer.	希望您能重新考慮。
Please take the advice into consideration.	請考慮此建議。
We look forward to your favorable reply.	我們期待您的回覆。
We look forward to your order.	我們期待您的訂購。
We hope you would place an order for this revised offer.	我們期待您能接受修改後的報價，並來信訂購商品。

實力大補帖 Let's learn more!

文法王 文法 / 句型解構

談判與議價

1 Let me explain to you why. 我來解釋一下原因。

2 That will eat up a lot of time. 那會耗費很多時間。

3 It depends on how much you purchase. 要看您的購買量而定。

4 Let's compromise. 我們各退一步吧。

訂購及回覆
Making a Purchase

5-01 訂購產品

訂購信件要以清楚為首要考量，因此建議用「條列式」的方法編寫。記得要在訂貨信件中清楚說明項目、金額、數量以及其他條件，以便對方處理訂單。

From:	Fiona Hall
To:	yang.ABCshoe@mail.net
Subject:	Place an Order

Dear Mr. Yang:

With **reference**[1] to your quotation of July 11, we would like to place a **firm**[2] order for:

(1) 250 pairs of men's leather shoes

Reference: 22245; size: 11 @$10 (inclusive of tax)

(2) 150 pairs of women's leather shoes

Reference: 23115; size: 6 @$12 (inclusive of tax)

Total (inclusive of tax): $4,300

Invoices[3]: 3 copies

Cash on **delivery**[4].

Please deliver within 3 weeks to our office in Taipei, and please **reply**[5] to confirm this order.

Yours sincerely,

Fiona Hall

楊先生您好：

根據您七月十一日的報價，我們希望正式訂購以下商品：

(1) 二百五十雙男用皮鞋

型號：22245；尺寸：11；每雙 10 元美金（含稅）

(2) 一百五十雙女用皮鞋

型號：23115；尺寸：6；每雙 12 元美金（含稅）

總價四千三百元（含稅），請附上三份發票，貨到付款。

請於三週內寄送至台北辦公室，並請回覆此信以確認訂購。

費歐娜・霍爾 謹上

key words ▲ E-mail 關鍵字一眼就通

1 reference 名 參考　　**2** firm 形 堅定的　　**3** invoice 名 發票

4 delivery 名 交貨　　**5** reply 動 回覆

🖎 複製 / 貼上萬用句 Copy & Paste

同意對方的報價

Thank you for your quotation of May 21 of the leather shoes.	謝謝您於五月二十一日寄來的皮鞋報價單。
We understand the price quoted is CIF Taipei.	我們了解您的價格為台北到岸價。
We accept your offer on these goods.	我們願意接受此報價。
We agree with your quotation.	我們同意您的報價。
We agree to your terms of payment and shipment.	我們同意您提出的付款條件以及送貨方式。

We have decided to accept the 15% discount you offered and terms of payment.	我們決定接受您所提供的八五折優惠以及付款條件。
All the terms you offered in your mail are agreeable to us.	我方同意您提供的所有條件。

下試用訂單

You have convinced me to place a trial order by your mail on May 11.	您於五月十一號的來信讓我決定下試用訂單。
We are disappointed with the bad discount; however, we will still place a trial order.	我們對於折扣不多感到失望，但仍會下試用訂單。
This is a trial order.	這是試用訂單。
If we are satisfied with your goods, we will certainly place further orders soon.	如滿意您的產品，我們會在近期內追加訂單。
If the goods sell well, we would place more orders in the near future.	若產品的銷售成績良好，我們會在近期內下更多訂單。
If this order turns out satisfactory, we would like to place a large order soon.	若此次的交易令人滿意，我們會在近日內下大筆訂單。

正式訂購

Having looked over your samples, we found that the quality of your goods meets our requirement.	看過貴公司的樣品後，我們認為您的產品符合我方對品質的要求。
As your goods are very popular in the market, we would like to place weekly orders with you.	由於您的商品在市場上很受歡迎，我們決定每週下訂單。

Please confirm that the quality of the goods is the same as that of the samples received.	請確認訂購貨物的品質與樣品一致。
Thank you for your catalogue concerning leather shoes; attached please find our order No.123.	謝謝您寄來的皮鞋目錄，在此附上我們編號 123 號的訂購單。
We are pleased to place a formal order for these goods.	我們正式向貴公司訂購此批貨物。
We would like to order the following items:	我們欲訂購以下商品：
Please provide the following goods:	請提供以下商品：
We would like you to send us the quantities of goods as follows:	請貴公司寄送以下數量的產品給我們：
We have studied your catalogue and confirmed our order for item No.101.	我們已看過目錄，並確認要訂購編號 101 號的產品。
We would like to request a further order by air as soon as possible.	我們想要追加訂單，並希望能及早空運。

提醒交貨時間

It would be highly appreciated for delivery by the end of June.	若能於六月底前交貨，我們將感激不盡。
We hope the products will be shipped by September 21.	希望貨品能在九月二十一日前寄送。
Please confirm that you can supply this quantity by the end of May.	請您確認五月底前可以提供此數量的貨品。
Please do your best to meet the delivery date of Nov. 22.	請您務必在十一月二十二日前出貨。

Please dispatch all items no later than October 20, 2018.	請於二〇一八年十月二十日前寄送貨物。
This order will be delivered within two weeks as guaranteed.	這份訂單條件為保證兩週內交貨。
These goods are in urgent need; please arrange immediate delivery from ready goods.	我們亟需此批貨品,請立即安排出貨事宜。
We will pay for urgent delivery fee.	我們會付急件的運費。
The items are urgently required; please arrange production as soon as possible.	我們亟需此批貨品,希望您能盡快安排上線生產。
We would like to receive these items as soon as possible.	我們希望能盡快收到此批貨物。
Please inform us if the expedited shipping is not possible.	如果商品無法快速送達,請通知我們。
We would like to ask you to ship the goods by FedEx.	我們希望您能以聯邦快遞寄出貨品。
Please ship the items as soon as possible via UPS expedited shipping.	請使用優比速快遞儘速寄出產品。

付款方式

Please inform us the payment method by return.	回函請告知付款方式。
Would you please let us know the shipping costs?	能否告知運費是多少呢?
Can we pay in NT dollars?	能否使用新台幣付款呢?
Please inform me if you cannot accept orders on a COD basis.	若您不接受貨到付款這種交易方式的話,請通知我。

The payment would be made in cash within ten business days.

本公司十個工作天內會付現。

The payment would be made quarterly.

我們將按季結算貨款。

Payment will be made by banker's draft tomorrow.

貨款明日將由銀行匯票支付。

寄送貨品 & 發票

Do you provide overseas delivery?

您的貨品有寄送海外嗎？

Please issue us a P/I (proforma invoice) for the goods.

請提供預付發票。

Please send us the original invoice plus five copies.

請寄送發貨單正本以及五份副本。

Please send the invoice to the address below:

請將發票寄送到以下地址：

Please attach the invoice, packing list and a B/L to the goods if you send them by air.

若使用空運，請將發貨單、裝箱單以及提貨單一併附在產品中。

Please send us by registered mail with invoice and packing list.

請將發貨單以及裝箱單以掛號寄出。

If any items are out of stock, please provide another quotation for a substitute.

若有商品缺貨，請提供替代品的報價單。

If you don't have them in stock, please do not send substitutes to replace them.

若商品缺貨，請勿寄送替代產品。

If you don't have these items in inventory, please replace them with another model of similar quality.

若此款商品缺貨，請提供品質相仿的其他產品。

Every item should be packed in an individual hard cardboard box.

每樣貨品都必須使用耐用紙盒包裝。

信件結語：請對方回函確認

We would be grateful if you could confirm the order by return.	如您能回函確認，我們將感激不盡。
We would like to ask you to send your order confirmation by return.	請您回覆，以確認訂購。
Please confirm this order by e-mail or fax.	請寄電子郵件給我們或傳真確認此訂單。
Your immediate attention would be highly obliged.	我們感激您的密切關注。

 實力大補帖 Let's learn more!

深入學 資訊深度追蹤

📍 訂購相關縮寫

在訂單往來的信件上經常會以縮寫表示，如果看不懂縮寫，那可就真的太不方便了。以下提示幾個常用縮寫所表示的英文，有商業往來需求的人請務必記住，才不會誤解對方的意思。

01 B/L = bill of lading 提貨單

02 P/I = proforma invoice 預付發票

03 P/O = purchase order 採購訂單

04 POD = proof of delivery 到貨證明

05 COD = cash on delivery 貨到付款

06 ETA = estimated time of arrival 預計到貨時間

Part
1
入門篇

Part
2
求職篇

Part
3
客戶往來篇

Part
4
人際互動篇

5-02 接受訂貨

　　接受訂貨時，須仔細確認所有的交易內容。國際貿易實務中，賣方會開出預付發票（Proforma Invoice），註明產品明細、交貨期限、付款條件、價格、數量等細節，並請客戶簽名確認。

From:	Terry Yang
To:	Fiona Hall
Subject:	Thank You for Your Order No.245

Dear Fiona:

I am writing to confirm your order via e-mail on July 20 for the following items:

(1) 250 pairs of men's leather shoes

Reference: 22245; size: 11 @$10 (inclusive of tax)

(2) 150 pairs of women's leather shoes

Reference: 23115; size: 6 @$12 (inclusive of tax)

As all the items are in stock, we would deliver them by UPS after receiving the payment.

We hope you will find our goods satisfactory. Once again, thank you for doing business with us and we hope to receive further orders from you.

Sincerely yours,

Terry Yang

費歐娜您好：

此函為確認您七月二十日所下的訂單，訂購明細如下：

(1) 二百五十雙男用皮鞋

型號：22245；尺寸：11；每雙 10 元美金 (含稅)

(2) 一百五十雙女用皮鞋

型號：23115；尺寸：6；每雙 12 元美金 (含稅)

產品皆有存貨，一收到貨款即會以優必速快遞寄出。

希望您滿意我們的產品。再次感謝您的訂購，也希望能夠與您再次合作。

楊泰瑞 謹上

複製 / 貼上萬用句 Copy & Paste

收到訂單的回覆

Thank you for your purchase order No.123.	感謝您來信訂購，訂單號碼 123 號。
Welcome to be our customer.	歡迎您成為我們的客戶。
We are pleased to acknowledge your order for these items below:	我們已確認您的訂單，如下：
Attached you will find your purchase information.	附件為您的訂購資料。
We ensure your order will have our careful attention.	我們保證會細心處理您的訂單。
We would arrange your order by this Friday.	本週五前我們會處理好您的訂單。
We are pleased to confirm your order as shown in the attached proforma invoice.	我們附上預付發票，以確認您的訂單資料。
Please open an L/C as soon as possible.	請及早開出信用狀。
Your order has already been arranged.	您的訂單已處理完畢。
We are now making arrangement for shipment.	我們目前正在安排貨運。

寄送貨品的時間

We have sent you some sales promotional literatures by airmail.	我們已將銷售文宣寄送給您。
Once we receive the payment, the goods will be dispatched immediately.	收到您的貨款後,我們會馬上寄出貨品。
Your order will be dispatched as soon as you complete the credit card information on our website.	您在網頁上填寫完信用卡資料後,產品便會寄出。
Delivery will be made immediately on receipt of your check.	收到您的支票後,產品會馬上寄出。
As soon as the order is ready, we will deliver them via FedEx.	貨品確認後,便會由聯邦快遞寄出。
All the items are in stock and can be dispatched to you by the end of May.	所有產品皆有庫存,並可於五月底前寄出。
Shipment will be executed by the end of June.	六月底前會出貨。
Delivery will be made by our own transport next Monday.	下週一會由我們自己的貨運單位寄送產品。
We are now arranging shipment in accordance with the terms agreed upon.	我們會根據協議的條件安排貨運。
The delivery will be made within 20 days after receiving your order.	收到訂單後的二十天內會寄出貨品。
As soon as we receive your L/C, we will ship the goods immediately.	收到您的信用狀後,我們會立即寄出貨品。
The order is guaranteed to be delivered to Taipei before July 13.	我們保證,商品會在七月十三日前寄達台北。
As you requested, we will inform you of the date of delivery immediately upon completing shipment.	我們會依據您的要求,於商品寄出後通知您出貨日期。

其他方面的配合

The order will be packed and delivered according to your instructions.	包裝方式和貨運細節，我們都會按您的要求安排。
These goods have been carefully packed in hard cardboard boxes.	產品已使用耐用厚紙盒包裝。
After you got the L/C, you can fax us first.	當您收到信用狀後，您可以先傳真一份給我們。
We would prepare materials and schedule your order earlier.	我們可及早安排產品以及運送時程。
As requested, invoice and packing list will be sent by registered mail.	我們已按您的要求，用掛號寄出發票及裝箱單了。

付款事宜

Attached are our payment forms, please complete the form and send it back.	附件為付款表格，請填妥後寄回。
You could contact us for other payment options.	若您想要詢問其他付款方式，可與我們聯繫。
We appreciated your generous orders recently.	感謝您近日大量訂購產品。
We would like to extend your payment deadline until the last day of June.	我們願意延長付款期限到六月三十日。
Please confirm by signing the P/I return.	請確認並簽妥預付發票後寄回。

信件結語

We sincerely appreciate your business with us.	我們很感激與貴公司的生意往來。

We hope our goods meet your expectations.	希望您滿意我們的產品。
We are sure that you will be extremely satisfied with our products.	相信您會非常滿意本公司的產品。
We hope your first order with us will lead to further business.	希望首次交易能夠帶來更多的合作商機。
Don't hesitate to let us know if you have any questions.	如有任何疑問，請隨時告知。
If you have any questions about the products, please contact me.	如您對產品有任何疑問，請與我聯繫。
Thank you again for giving us the chance to serve you.	再次感謝您給予我們服務機會。

實力大補帖 Let's learn more!

活字典 單字 / 片語集中站

📍 確認訂單

acknowledge 告知收到	procurement 採購
inventory 存貨	in stock 有現貨
out of stock 售罄	a rush order 緊急訂單
process the order 處理訂單	fulfill the order 執行訂單

📍 寄送貨品

dispatch 快遞	transport 運輸
shipment 運送	free shipping 免運費
shipping fee 運費	in-store pickup 到店取貨

在「拒絕接受買方訂購」的信件中，須禮貌並誠懇地表達無法接受訂單的原因，並提供其它可行的替代方案，以製造下一次的合作機會。謹記，切勿輕易喪失與客戶合作的任何可能性。

From:	Nancy Huang
To:	Mr. Jones
Subject:	Declining Order

Dear Mr. Jones:

We thank you very much for your order No.123 of June 12. **Unfortunately**[1], we are very sorry that we will not be able to **fulfill**[2] your order **due to**[3] leather **materials**[4] **shortage**[5]. In fact, the leather shoes are out of stock now, and we are now unable to order.

However, we will try to order the materials from abroad, but there will be a delay of up to 5 weeks for production **cycle**[6]. We would appreciate if you agree to this.

Looking forward to your reply.

Yours truly,

Nancy Huang

瓊斯先生您好：

謝謝您於六月十二日寄來的 123 號訂單。很遺憾地，目前由於皮革原料短缺，導致皮鞋製品缺貨，因此我們無法處理您的訂單，對此我們感到非常抱歉。

我們會試著從國外訂購原料，但由於製作流程的關係，所以出貨會延遲（最遲會晚五個禮拜）。若您能接受，我們將會非常感激。

期待您的回信。

黃南西 敬上

key words E-mail 關鍵字一眼就通

1 unfortunately 副 遺憾地　2 fulfill 動 滿足　3 due to 片 因為

4 material 名 原料　5 shortage 名 短缺　6 cycle 名 整個過程

 複製 / 貼上萬用句 **Copy & Paste**

拒絕訂單

We regret to inform you that we cannot accept such a small order quantity.	很遺憾，我們無法接受這麼小額的訂單。
We are sorry that we must turn down your order.	我們很遺憾必須拒絕您的訂單。
We are unable to comply with your request due to time constraints.	由於時間不足，我們無法處理您的訂單。
Regretfully, we have to decline your order as our factory will be closed for two months.	很遺憾地，因為工廠休息兩個月，所以我們必須婉拒您的訂單。
Chinese New Year holidays are coming; therefore, the workers will take their vacation from the end of January.	農曆新年即將到來，員工將從一月底起開始休假。
That is the reason for us to decline your order.	此為拒絕您訂單的原因。
The best delivery time we can make is by the end of September.	我們最快也要到九月底才能出貨。

付款與價格的爭議

According to the terms of payment, we regret we cannot accept the D/P terms.	根據付款條件，我們無法接受付款交單。

We are sorry that we cannot grant you a draft at 60 days under D/A.	很抱歉我們無法接受六十天承兌交單之匯票。
We only accept the order by cash.	我們只接受付現的訂單。
We only accept an irrevocable L/C at sight or T/T wired before shipment.	我們只接受在貨運前收到不可撤回之信用狀或電匯單。
It is our company policy.	此為公司政策。
If the revised quotation is unacceptable, we have no choice but to hold this.	若仍無法接受修正過後的報價，那我們只好暫時擱置此訂單。

缺貨與斷貨

We are sorry to inform you that the items mentioned in this order are currently out of stock.	我們很遺憾告知：您要的商品目前缺貨。
There is no inventory in our warehouse now.	目前倉庫無存貨。
Due to recent increase in sales, we don't have any inventory in stock.	由於近日的銷售量增加，所以目前已無存貨。
We are now short of stock.	目前存貨短缺。
The delivery date requested by you does not grant us sufficient time for preparation.	貨品需要的準備時間趕不上您所要求的出貨日期。
Our production lines have been fully booked now.	我們的生產線已滿單。
The production line is all booked until the end of September.	截至九月底的生產線已滿單。
The product of your order has been EOL.	您所需的貨品已經斷貨了。

Part
1
入門篇

Part
2
求職篇

Part
3
客戶往來篇

Part
4
人際互動篇

Production of that special model has ended.

該項特殊產品已經停產了。

Our company has ceased production of this model since last year.

我們已在去年停止生產此項商品。

We no longer manufacture this product over two years.

我們已經停止生產此項商品兩年多了。

因各項原因延遲

Due to delivery issue, there will be a delay of up to 8 weeks.

由於貨運問題,將會有八週左右的延遲。

Due to supply shortage in China, your order will be delayed.

由於中國方面的供應短缺,您的貨品將會延遲出貨。

The minimum period required for preparation is 6-8 weeks.

準備時間最少需要六至八週。

The minimum period for production of these items is 4 weeks at least.

製造這些商品的時間至少要四週。

We hope you would give us more time for stock and shipment preparation.

我們希望您能多給我們一些時間準備商品以及貨運。

提供替代方案

We can supply you with other similar products from the stock.

我們可提供您其他類似的產品。

We put forward our new model 1306 for your consideration, which can be a good replacement.

請考慮以我們型號 1306 號的新產品做為替代。

The function and specifications are greatly improved.

其功能與規格皆改良過。

The suggested item sells well in the market.

此建議商品在市場上的銷售表現良好。

We recommend other models with similar functions and quality.	我們推薦您購買其他功能及品質相仿的產品。
If the suggested model meets your requirement, please inform us soon.	如建議的產品符合您的需求，請儘速告知我方。
We are confident the new model will give you complete satisfaction.	我們有信心您會非常滿意這項新產品。

信件結尾：致歉

We apologize for not being able to accept your order.	我們對於無法接受您的訂單感到抱歉。
We are really sorry for declining your order.	有關拒絕訂單一事，我們深感抱歉。
If you would like to cancel this order, it would be understandable.	若您決定取消訂單，我們能夠諒解。
Again, we extend our sincerest apologies and hope you can understand.	再次祈求您接受我們的致歉。
Under the circumstances, we are sorry that we have to turn down your order.	在此狀況下，我們必須婉拒您的訂單。
In this case, it would be better for us to decline your order.	在此狀況下，我們必須婉拒您的訂單。
We hope you will understand our difficulties and situation.	希望您能了解我們的困難和處境。
Please inform us your decision as soon as possible.	請儘早通知我們您的決定。
We hope for other opportunities to do business with you.	我們希望仍有其他與您合作的機會。
We hope we can serve you in the future.	我們希望未來能替您服務。

5-04 取消訂單

造成取消訂單的原因很多，如果是因為對方的疏失，可展現強硬的態度；如果是本身的決定，則宜簡短、清楚地說明取消訂單的理由，但結尾仍須以謙虛有禮的態度請求對方諒解，以維持雙方愉快的合作關係。

From:	Betty Chang
To:	Mr. Wu
Subject:	Cancellation of Order

Dear Mr. Wu:

We have placed our order for 500 **electronic**[1] **calculators**[2]. However, we have to **cancel**[3] this order due to customer **de-commit**[4].

We hope you will agree to cancel this order, sorry for the **inconvenience**[5].

Thank you **in advance**[6].

Yours truly,

Betty Chang

吳先生您好：

我們本來有向貴公司下一份五百台計算機的訂單，然而，因客戶取消訂購，我們也必須取消訂單。我們希望您能同意取消，造成不便我們深感抱歉。

在此感謝您。

張貝蒂 謹上

key words ▲ E-mail 關鍵字一眼就通

1 electronic 形 電子的　　**2** calculator 名 計算機　　**3** cancel 動 取消

4 de-commit 名 取消；撤回　**5** inconvenience 名 不便　**6** in advance 片 預先

取消訂單的理由

🖱 Due to financial reasons, we have to cancel our order.	因為財務上的考量，我們必須取消訂單。
🖱 Due to customer cancellation, we are forced to cancel this order.	由於客戶取消訂購，我們被迫取消訂單。
🖱 Due to the bad quality of your products, we regret that we have to cancel our order.	由於產品的品質不佳，我們很遺憾地必須取消訂單。
🖱 We have stressed the importance of shipping our order by the end of May.	我們已強調貨運必須在五月底前送達的重要性。
🖱 We have to cancel this order since we have not received the goods yet.	因尚未收到任何貨品，我們必須取消訂單。
🖱 Due to your delay in shipment, we decided to cancel the order.	因貨運延遲，我們決定取消訂單。
🖱 Please cancel the order because our customer cannot accept the delayed shipping date.	由於我們的客戶無法接受如此延遲的送貨日期，因此，我們必須取消訂單。

信件結語：致歉

🖱 We are very sorry for the inconvenience.	很抱歉造成不便。
🖱 We hope you will accept our sincere apologies.	希望您接受我們誠摯的道歉。
🖱 Please accept our sincere apology.	請接受我們誠摯的道歉。
🖱 Thank you so much for your understanding.	謝謝您的諒解。

5-05 催促買方訂貨

於「催促買方訂貨」的信件中，應給予催促的原因，例如價格即將調整、或是存貨即將售罄，以增加買家想要儘速下訂單的動機。

From:	Rita Moore
To:	Mr. Freeman
Subject:	Re: Item No.6668

Dear Mr. Freeman:

Please refer to our e-mail dated August 23 and our quotation for item No.6668. We would like to know if you have any conclusion now, and can you confirm the order as early as possible?

Due to the rising cost of raw materials, the price of our products will go up soon. However, we will maintain our offer for you until the end of September.

Please understand this situation, and take this opportunity by confirming your order soon.

Truly yours,

Rita Moore

費曼先生您好：

請詳見八月二十三日信中，我們針對編號 6668 號貨品的報價。我們想要了解您是否有了結論，可否請您儘速確認訂單呢？

由於原物料上漲的關係，我們的產品也即將漲價。然而，我們會維持此報價到九月底。

感謝您的諒解，若要下單的話，還請儘早確認。

芮塔‧摩爾 敬上

訂單量的考量

Due to the sudden increase of orders, we would like to confirm with you about your order.	由於訂單量激增，我們想要請您確認訂單。
These products are very popular in Taiwan; therefore, we have received a large amount of orders.	這些產品在台灣非常受歡迎，因此我們收到大量訂單。
As we have got many orders this month, it would be better for you to place an order as early as possible.	由於本月的訂單量很大，還請您及早下訂單。
As the demand will be very huge before Christmas season, we hope you will place an order as early as possible.	由於聖誕節前夕的產品需求量非常大，我們希望您能儘早下訂單。
If you wish to ensure delivery before Chinese New Year, please place an order soon.	若您希望在農曆年前收到貨品，請及早下訂單。
We would suggest you give us your order by this weekend since our schedule is very tight.	由於我們的生產線非常忙碌，希望您能於本週末前下訂單。
As our schedule is booked up until September, please confirm your order soon.	生產線目前已滿單至九月底，請您儘速確認訂單。
Our production line is very full; therefore, we advise you to place an order soon.	我們的生產線非常忙碌，建議您及早下訂單。
We urgently need to know the quantity of your order so that we can arrange production for you.	我們亟需知道您的訂單數量，以便為您安排生產。

We would like to arrange production for you.

我們想要為您安排生產線。

We would like to put your order into production this month.

我們想要為您安排本月的生產日程。

價格恐調漲

It's the best quotation we can offer; we would advise you to place your order as soon as possible.

這是我們所能提供的最優惠報價，我們建議您儘速下單。

As you know, the price of raw materials has been rising these two years.

如同您也了解的，這兩年原物料的價格上漲。

We will raise our prices after the present stocks are cleaned out.

此批庫存售罄後，我們將會漲價。

As soon as we run out our stocks, the price will be revised.

庫存售罄後，價格將會調整。

產品的特殊之處

Our products are superior to many similar ones on the market.

我方產品比起市場上相仿的產品優良許多。

With fine quality and attractive design, our goods have been selling extremely well.

我們公司產品的品質好、設計佳，因此賣得極好。

Our terms and prices are very competitive.

我方所開的條件以及價位都非常有競爭力。

Our price is extremely low for this excellent quality.

以這麼優良的品質而言，我方的價位已經非常低了。

We are sure it will bring you profits if you place orders regularly.

如果您定期下訂單的話，我方的產品一定會為您帶來利潤。

After our current stock run out, we would not manufacture the same model in the future.

目前的庫存售完後，我們就不會再生產這批型號的產品了。

信件結語

We would suggest you to seize the chance and place an order.

我們建議您把握時機下單。

We hope you will place an order before price hike.

希望您在我們漲價前下單。

We would be delighted if you place an order with us.

若您能與我方合作，我們將會非常高興。

實力大補帖 Let's learn more!

文法王 文法 / 句型解構

● as long as / as soon as / as far as 的用法解析

1 as long as 只要…（連接詞）。表示條件，用法相當於 if。置於句首的 as long as 需要加上逗號；置於句中則不用。

ex. We are happy to place an order as long as you give us a discount.
只要您降價，我們很願意下單。

2 as soon as 一…就…（連接詞）。用於連接「同一時刻發生的兩個動作」。置於句首的 as soon as 需要加上逗號；置於句中則不用。

ex. As soon as you confirm the price, we will inform our customer.
一旦您確認價格後，我們就會通知客戶。

3 as far as 則有兩種用法，第一種表示「實際的距離有多遠」；第二種則被當作「連接詞」使用，表示「就…而言」。

ex. As far as I'm concerned, the price they offer is very competitive.
我認為他們出的價格很有競爭力。

6-01 要求對方提供信用狀

　　信用狀（Letter of Credit）為銀行根據進口商的指示向出口商開出的文書，出口商提供合乎規定的單據和匯票，銀行即可付款。貨物即將安排裝運時，也需要信用狀的確認來安排船期。

From:	Stella Wilson
To:	Mr. Pan
Subject:	L/C for Order No.5556

Dear Mr. Pan:

We thank you for your order No.5556 of March 20 with which you have sent us the shipping instructions.

We would like to ask you to promptly open an L/C in our favor and fax us the copy of it first. Therefore, we can start to arrange shipment immediately.

We thank you in advance.

Best regards,

Stella Wilson

潘先生您好：

感謝您三月二十日來函下訂單 5556 號，並告知貨物的運送指示。

我們希望您能儘速開立信用狀，並先將影本傳真過來，我們才能開始準備貨運。

先跟您說聲謝謝。

史黛拉‧威爾森 謹上

複製 / 貼上萬用句 Copy & Paste

信件開頭 & 信用狀

Thank you for your mail containing your acceptance of our offer.

謝謝您來信接受我們的報價。

We ask that you open an L/C in our favor, valid until May 31.

請您開出有效期至五月三十一日的信用狀。

According to the policy of our company, shipment is to be arranged after receipt of the L/C.

根據本公司政策，收到信用狀後，我方才會準備貨運。

We will ship your goods as soon as we receive your L/C.

一接到信用狀，我們就會準備送貨品。

Would you please inform us when you expect to open the L/C?

可否請您回覆將於何時開立信用狀？

We are awaiting the confirmation of the L/C.

我們正在等待對方開立信用狀。

較為急切的說法

You have mentioned in the mail that you had notified your bank to open the L/C.

您的來信中提到，您已經請銀行開立信用狀。

We haven't received your L/C yet.

我們尚未收到信用狀。

We need you to open the L/C urgently in order not to make shipment delay.

我們需要您盡快開立信用狀，以免延遲貨運時程。

Without the L/C, we cannot begin production.

如果沒有信用狀，我們無法安排生產。

Unless the L/C is opened in time, we will quote you a new price.

如果信用狀無法如期開出，我方會另外開價。

Part
1
入門篇

Part
2
求職篇

Part
3
客戶往來篇

Part
4
人際互動篇

🖱	Upon receiving confirmation of the L/C, we will promptly pack and ship the goods.	收到信用狀後，我們會馬上包裝出貨。
🖱	Upon receiving your L/C, we will complete the shipment of your order.	收到信用狀後，我們就會出貨。
🖱	As soon as we received confirmation that the L/C has been established, the goods will be shipped.	信用狀確認開出後，我們便會寄出貨品。
🖱	Upon arrival of the L/C, we will pack and ship the products urgently as requested in accordance with your shipping instructions.	一收到信用狀，我們便會依照您的貨運指示盡快包裝以及運送。
🖱	We assure we will complete shipment and give you perfect satisfaction.	我們會完整的裝運，一定會讓您滿意的。
🖱	Please take this as an urgent issue.	請儘速處理此事。
🖱	Please confirm it as soon as possible.	請儘速確認。

🛩 實力大補帖 Let's learn more!

深入學 ▲ 資訊深度追蹤

📍 何謂信用狀 L/C ？

　　信用狀為買方委託銀行對賣方開出的一種文件，銀行藉此向賣方保證，在其履行約定條件後，將對賣方之匯票予以承兌或付款；可說是開狀銀行開給賣方的付款保證書。

　　使用信用狀對買方及賣方的優點為保障雙方信用、確定契約順利完成、以及資金運用便利；對進口商而言，也可藉由信用狀確定履行契約的時間，因此，信用狀在國際貿易中經常被使用。

6-02 通知信用狀已開出

在「通知信用狀已開出」的信件中，告知對方開立信用狀的銀行名稱、金額、與開立日期，並須表明拿到信用狀後，會立即傳真影本給對方確認。

From:	Jack White
To:	Ms. Hill
Subject:	L/C Confirmation - No.3365

Dear Ms. Hill:

Please be informed we have asked our bank, Bank of Taiwan, to open the L/C for $50,000 covering our order No.3365. The L/C will be opened to you by this Friday.

As soon as we get the L/C details, we will fax you a copy immediately.

Best regards,

Jack White

希爾女士您好：

請知悉我們已經請台灣銀行針對我方 3365 號訂單開立五萬美元的信用狀。信用狀將於本週五前開出。

我們一拿到信用狀，就會馬上傳真影本給您。

傑克‧懷特 敬上

複製 / 貼上萬用句 Copy & Paste

信用狀相關

We are sorry for the delay to open the L/C.	我們很抱歉延遲開立信用狀。
We have instructed the Bank of Taiwan to open an L/C for $6,500.	我們已經指示台灣銀行開立六千五百美元之信用狀。
In reply to your e-mail of March 12, we have already opened the L/C.	回覆您三月十二日的來信，我們已開出信用狀。
The letter of credit will be valid until November 25.	信用狀的期限為十一月二十五日。
The letter of credit will be sent to our banker's correspondent, Barkleys Bank in England.	信用狀會寄送到我們在英國的通訊銀行—巴克萊銀行。
The letter of credit will cover your invoice for the amount of the goods.	信用狀金額會涵蓋產品的發票金額。
Once we have received the L/C from our bank, we will fax you a copy of it.	我們一拿到信用狀，就會立即傳真影本給您。

信件結語：請對方確認

Please be informed that we have opened the L/C for $10,000 covering our order.	請確認我們已根據訂單開立一萬美元之信用狀。
Please check with your bank.	請與您的銀行確認。
Please arrange the shipment as soon as possible.	請儘速安排貨運。
Please carry out our order as early as possible.	請儘速處理我們的訂單。

出口商一旦發現信用狀需要修改時，應立即向客戶反應，並於信中清楚表達需修改的內容，或是在信中註明原本錯誤的內容與修改後的內容，以確保自己的權益，並避免延誤交貨時間。常見的修改事項有：受益人的名稱或地址、延長船期、延長信用狀有效期限、金額或貨品的數量有誤、交貨期限不正確等。

From:	Tina Chen
To:	Mr. Brown
Subject:	L/C No.123 Amendment

Dear Mr. Brown:

We would like to explain the matter of L/C No.123 **amendment**[1].

First of all, thank you very much for the L/C you sent yesterday. But we are sorry to inform you that we need you to amend the following points:

(1) **Extend**[2] the shipping date to June 30, 2018.

(2) Extend the **expiry**[3] date to July 15, 2018.

As our **production**[4] line has been fully **booked**[5] in May, the earliest **shipment**[6] we can assure you will be around the end of June instead of May. Please amend it as soon as possible.

We look forward to receiving the L/C amendment soon.

Best regards,

Tina Chen

布朗先生您好：

我們欲解釋 123 號信用狀的修改事宜。

首先，感謝您昨日寄來的信用狀。很抱歉，我們需要針對以下幾點做更改：

一、貨品運送日延至二〇一八年六月三十日。

二、延長有效日期至二〇一八年七月十五日。

由於五月份的生產線繁忙，我們可以保證的最早出貨時間為六月，而非五月，煩請儘早修改。

我們盼望能早日收到修改過後的信用狀。

陳蒂娜 敬上

E-mail 關鍵字一眼就通

1 amendment 名 修正　　2 extend 動 延長　　3 expiry 名 期滿

4 production 名 生產　　5 book 動 預訂　　6 shipment 名 運送

複製／貼上萬用句 Copy & Paste

點出信用狀有誤之處

Thank you for your cooperation in opening L/C, which we have just received this morning.	謝謝您開立的信用狀，我們已於今早收到。
Your goods are ready, but we cannot ship them due to errors in the L/C.	您的貨品已經準備好，但因為信用狀有誤而無法寄出。
Therefore, the L/C needs urgent amendment.	因此，信用狀必須立即更正。
With reference to your L/C No.12345 covering the order No.5556, we regret to inform you that the factory will be closed for one month due to the Chinese New Year vacation.	根據您 5556 號訂單之 12345 號信用狀，我們必須通知您：適逢農曆新年，因此工廠將會休息一個月。
There seems to be an error in the quantity shown on the L/C.	信用狀上的數量似乎有誤。
The exact quantity is 5,225 pieces, not 5,252 pieces.	正確數量為五千二百二十五件，而非五千二百五十二件。

Due to peak season, it's very difficult to book shipping schedule.	由於正值旺季，船期非常難安排。
Due to full production schedule, it's difficult for us to ship the goods before Chinese New Year.	因為生產線繁忙，我們認為於新年前出貨有困難。
We would appreciate if you comply with our request and extend the shipping date to June 30.	如果您能依照我們的要求延長船期至六月三十日，我們將非常感激。
We would be obliged if you would extend the shipping date to September 15 and the expiry date to October 15.	如果您能延長船期至九月十五日、以及有效期限至十月十五日，將非常感激。
We want the payment to be made at sight.	我方希望見票即付。
Please amend trade terms to be on FOB basis instead of CIF.	請將到岸價格更正為離岸價格。

信件結語

We would appreciate your immediate amendment of the L/C.	希望您能盡快更改信用狀。
We must ask you to amend the L/C.	我們必須請您修改信用狀。
We look forward to receiving your L/C amendment ASAP.	我們盼望能盡快收到修正後的信用狀。
Your prompt attention to this matter would be much appreciated.	我們將非常感激您的儘速處理。

6-04 信用調查（向備詢銀行查詢）

信用交易有時需要蒐集信用資料，此信為向第三者銀行請求提供信用資料的信件。
（※ 注意：文後必須向收信者保證得到的金融資訊將會嚴格保密。）

From:	William Franco
To:	X&Y Bank
Subject:	Referring to the Reference

Dear Sir or Madam:

We have received an order with $70,000 for Anderson **Engineering**[1] Co. Ltd. in Taipei, and they provide us your name as a reference.

We would like to have some information about their **financial**[2] **standing**[3] before doing business with them.

Any information you send us will be held in **strict**[4] **confidence**[5]. We would appreciate any information you can provide.

Yours faithfully,

William Franco

敬啟者：

我們接到台北安德森工程有限公司七萬元的訂單，他們指定貴公司為備詢銀行。

在與該公司交易前，我們希望能夠了解一下該公司的財務狀況。

您所提供的資訊都將嚴格保密，我們會非常感激貴公司所提供的資料。

威廉‧法蘭科 敬上

 E-mail 關鍵字一眼就通

1 engineering 名 工程　　**2** financial 形 財政的　　**3** standing 名 狀態

4 strict 形 嚴格的　　**5** confidence 名 祕密

The Anderson Engineering Co. Ltd. has given us the name of your bank as a reference.	安德森工程有限公司指定貴公司為備詢銀行。
Before dealing with them, we would like to have some information about their financial standing and the range of their business.	交易前,我們希望能夠了解一下該公司的財務狀況及其經營範圍。
We would appreciate if you can give us their financial standing.	若貴公司能提供他們的財務狀況,我們會非常感謝。
Thank you for your help.	感謝您的幫忙。
An early comment would be highly obliged.	感謝您的及早回覆。

實力大補帖 Let's learn more!

深入學 資訊深度追蹤

信用調查之意義

　　國際貿易是一項具風險的事業,自買賣雙方開始接洽,經過詢問產品、寄樣品、報價、訂契約,至交貨、付款等,需要一段長時間的過程。為避免損失,在開拓外銷市場時,最好先予以信用調查(新客戶尤其該如此)。事前做信用調查,才能知曉對方公司的財務結構是否有什麼問題,合作也才能無虞。

信用調查之好處

　　信用調查可幫助公司選定信用良好的客戶與廠商。事前了解對方的信用程度,也好擬定給予賒帳的限額。即使是老顧客或廠商,也可以定期作信用調查,了解對方信用的動態,以防萬一。

6-05 信用調查（向備詢商號查詢）

此信為向公司或個人提出的信用諮詢，在信中簡介自己的公司，並說明意圖以及欲了解哪家公司的資料，如該公司之交易內容、額度與信用等，並保證不洩漏資料。

From:	Vicky Johnson
To:	Ms. Watson
Subject:	Referring to the Reference

Dear Ms. Watson:

We are the manufacturer of **kitchenware**[1]. Your name was given to us as a reference by Mr. Liao, one of our buyers, Otis Kitchen Supply, who has placed an order with us on 60-day **bill**[2] of exchange **basis**[3].

We would appreciate any information that could help us make the right decision. Please **specify**[4] the **period**[5] of your business relationship, and the **approximate**[6] volume of business done with you yearly.

We assure you that the information provided will be treated in strict confidence.

We are grateful for an early reply.

Yours faithfully,

Vicky Johnson

華生女士您好：

我們是廚房用品製造商。我們的買家──歐提斯廚具的廖先生要求給予六十天期匯票的付款方式，因此將您的名字提供給我們做為備詢。

若您願意提供任何能幫助我們做正確決定的資訊，將非常感激。煩請說明您與廖先生的交易時間長短，以及每年的平均交易額。

我們保證您所提供的資料將會嚴格保密。

感謝您的及早回覆。

薇琪‧強森 敬上

1 kitchenware 名 廚房用具　2 bill 名 匯票　3 basis 名 基礎

4 specify 動 詳細指明　5 period 名 期間　6 approximate 形 大約的

複製 / 貼上萬用句 Copy & Paste

信件開頭：緣由

Mr. Liao wishes to open an account with us and has provided your name as a reference.	廖先生希望能開立帳戶，並提供貴公司以茲備詢。
Anderson Company has purchased an order of $10,000 and has listed you as a credit reference.	安德森公司已下了一萬美元的訂單，並提供您為信用備詢人。
Mr. Chen told us you are prepared to act as their reference.	陳先生告知我們您可以擔任他們的信用備詢人。
The company stated they have business relationship with you for two years.	該公司告知我們與您有兩年的合作關係。

請對方提供資訊

We would be grateful if you can provide your view about the company's general standing.	若您能針對該公司的營運狀況提出看法，我們將非常感激。
We would like to know if the company is creditworthy.	我們想知道該公司是否值得信賴。
We would like to consult your opinion about whether the company is reliable for credit up to $20,000, and has the ability to settle their account promptly.	我們想了解該公司是否值得信賴，與是否有可承擔兩萬美元的信用額度帳戶。

We would like to know if the company is able to repay a loan of this amount within specified time.	我們想了解該公司是否有能力在期限內償還如此數量的貸款。
We would appreciate any information you provide on the credit history of Anderson Company with your company.	若能提供貴公司與安德森公司的交易紀錄，我們將感激不盡。
How long has the company been doing business with you, and how's its reputation for meeting obligations?	該公司與您的交易時間長短為何，以及他們的償債聲譽如何？
We would like to know if the owner has any outstanding debts.	我們想了解該負責人是否有未償清的債務。

信件結語：保證與感謝

Thank you for your assistance on this matter.	謝謝您於本事宜之協助。
We would appreciate any information you are able to offer in this condition.	若您能提供任何資料，我們將感激不盡。
We assure any information you send us will be kept confidential.	我們保證，您所提供的資料將會保密。

實力大補帖 Let's learn more!

活字典 單字／片語集中站

♀ 保密與洩密

be liable for 負有法律責任的	confidentiality 機密
proprietary 私有的；財產的	written record 書面紀錄
secrecy 祕密狀態	set forth 列舉
breach 違反	disclose 透露

6-06 信用調查的回覆（贊同）

提供贊同的信用資料時，應根據自家公司與該公司交易的情況給予說明，推測該公司值得信賴。另外，也可在信中特別註明，希望所提供的資料能保密。

From:	Lillian Lee
To:	Mr. Bloor
Subject:	Re: Referring to the Reference

Dear Mr. Bloor:

Thank you for your mail of May 25.

It is our **pleasure**[1] to inform you the company you **mentioned**[2] is a well-known **private**[3] company in our country. It has been **established**[4] in Taipei for more than 45 years.

We have had business relationship with them for 5 years on **quarterly**[5] account terms. Their **settlement**[6] has always been prompt. The credit we allow this company is also US$5,000 as you mentioned.

We hope the information is useful to you, and please contact us if you have any further inquiries.

Yours faithfully,

Lillian Lee

布魯爾先生您好：

謝謝您五月二十五日的來信。

我們很樂意告知您，您所提及的公司為一間在本國享有優良聲譽的私人公司，該公司在台北創立已超過四十五年了。

我們已有五年採用每季付款交易的往來紀錄，他們一向準時繳款。我們給予該公司的信用額度同樣也是五千美元。

希望此資料對您有幫助，如有任何疑問，請與我們聯繫。

莉莉安・李 謹上

key words ▲ E-mail 關鍵字一眼就通

1 pleasure 名 愉快	2 mention 動 提及	3 private 形 私營的
4 establish 動 建立	5 quarterly 形 按季度的	6 settlement 名 結帳

 複製 / 貼上萬用句 Copy & Paste

提供相關資訊

Thank you for your mail concerning the credit of our customer.	謝謝您來信諮詢我們客戶的信用情況。
We are pleased to provide a reference for the company.	我們很樂意擔任該公司的信用備詢參考人。
We have been dealing with the company for 20 years.	我們與該公司從事交易已有二十年。
The firm is a reputable one.	此公司的聲譽良好。
The firm has an excellent reputation in the trade.	此公司在商界有良好聲譽。
The company has a good reputation for conducting their business.	此公司擁有良好的經營口碑。
The owner is highly respected and is regarded as a business leader in this area.	該名負責人備受尊敬，並為本區的商業領導者。
We normally allow them credit facilities up to US$20,000.	通常我們提供兩萬美元的信用額度給他們。
The record of the payment is entirely satisfactory.	付款紀錄很令人滿意。

無法提供資訊

We are afraid that we cannot give you any information without the permission from the customer.	因為沒有得到客戶允許，我們恐怕無法提供您任何資料。
We know nothing about the company because we only deal with them on a cash basis.	我們不清楚該公司的狀況，因為我們都是以現金付款的方式交易。

信件結語：請對方保密

Please treat this information as confidential.	請將此資料保密。
We would remind you to consider the information confidential.	我們提醒您，請針對此資料進行保密。

實力大補帖 Let's learn more!

深入學 資訊深度追蹤

📍 針對新客戶所做的信用調查

在業務部門建立了潛在客戶的資料之後，審查部門就應針對客戶做調查，最普遍的信用調查包含下面的 5C：

01 品格（Character）：指客戶的信譽，愈有信譽表示對方履行償債義務的可能性愈高。

02 能力（Capacity）：指客戶的償債能力，此涉及對方的資產與負債比例。

03 資本（Capital）：指客戶的財務實力與狀況。

04 抵押的擔保品（Collateral）：當客戶無力償還債務時，能被用來當作抵押的資產。（當然，做信用調查時，還必須確認擔保品的價值。）

05 營業條件（Condition）：指可能影響客戶償債能力的情況。

Part
1
入門篇

Part
2
求職篇

Part
3
客戶往來篇

Part
4
人際互動篇

6-07 信用調查的回覆（不贊同）

提供不贊同的資料時，內容不宜過於露骨，簡單陳述事實即可，私人的主觀意見應予以保留，以不影響他人決定，也不損害該公司之聲譽為回覆的前提。

From:	George Harris
To:	Mr. Simpson
Subject:	Re: Referring to the Reference

Dear Mr. Simpson:

We have done business with the company you mentioned for many years. Although we believe the company is **reliable**[1], we have to **remind**[2] you that they have not always settled their **accounts**[3] **promptly**[4]. Furthermore, it's a private and small company. We think it's a case in which **caution**[5] is necessary.

We offer this information on the strict understanding that it will be treated **confidentially**[6].

Yours faithfully,

George Harris

辛普森先生您好：

我們已與您提及的公司交易多年。雖然我們認為此公司值得信賴，但我們必須提醒您，他們並非總是準時付款。另外，該公司為小規模的私人企業，我們認為有必要提醒您。

請務必將此份資料列為機密。

喬治・哈里斯 敬上

key words ▲ E-mail 關鍵字一眼就通

1 **reliable** 形 可信賴的　　2 **remind** 動 提醒　　3 **account** 名 帳目

4 **promptly** 副 迅速地　　5 **caution** 名 告誡　　6 **confidentially** 副 祕密地

複製 / 貼上萬用句 Copy & Paste

The company tended to delay payment for several times.	此公司曾數度延遲付款。
We have to remind them for the payment for many times.	我們需要多次提醒他們付款。
We never allowed them to reach the sum mentioned in your mail.	我們從未同意給予他們您信中提及的信用額度。
This company operates in a small scale.	此為小規模經營的公司。
It seems they have bad management and overtrading issues.	他們似乎有管理以及超額交易的問題。
This information is supplied in the strictest confidence.	這些資料請嚴加保密。

實力大補帖 Let's learn more!

活字典 單字 / 片語集中站

📍 債務與風險

current asset 流動資產	current liability 流動負債
deficit 赤字	debt 債務
debtor 負債者	deduction 扣除
draft 匯票	insolvent 無力償還的
liquidation 償付	mortgage 抵押
overdraft 透支	uncertainty 不確定性

7-01 出貨通知

From:	Bella Jones
To:	Mr. Hu
Subject:	Shipment (Order No.12553)

Dear Mr. Hu:

We have shipped your order No.12553 this morning. It will arrive at Taipei on the 20th at 5:00 p.m. If you have any questions, please let me know as soon as possible.

Thank you for doing business with our company.

Best regards,

Bella Jones

胡先生您好：

我們已於今早寄出您的 12553 號訂單商品，將於二十號下午五點抵達台北。若您有任何問題，請儘早告知。

感謝您與本公司交易。

貝拉‧瓊斯 敬上

複製 / 貼上萬用句 Copy & Paste

通知出貨

Thank you for doing business with our company.

謝謝您與本公司交易。

This is as a shipment notification.

此為出貨通知。

Please note that your order has been dispatched this morning.

您的訂單今早已出貨。

Your order No.335 has been shipped by Evergreen Shipping Co. Ltd. sailing on May 20 from Kaohsiung to Edinburgh.

您的 335 號訂單已由長榮海運自高雄運至愛丁堡。

We are pleased to inform you that your order has been dispatched.

我方樂於通知：您的貨品已寄出。

We shipped the following items on November 20, 2018.

下列物品已於二〇一八年十一月二十日寄出。

As promised, we had made the shipment on April 1.

如承諾，我方已於四月一日完成出貨。

As you requested, we have shipped your order by air on May 20.

如您要求，我們已將您的貨品於五月二十日空運寄出。

The items were shipped to the address you requested this afternoon.

商品已依照您指示的地點，於下午寄出了。

預估貨品抵達日

The estimated arrival time for your order will be on June 15.

您應該會在六月十五日左右收到貨品。

Your item will be shipped to the port tomorrow afternoon.

您的貨品將於明日下午送達港口。

It will take five to seven working days for the goods to arrive at the required address.	貨品需要五至七個工作天寄達您指定的地點。
As agreed, the products will be delivered to your Taipei office on Tuesday morning.	按雙方同意，貨品將於週二早上寄送至您位於台北的辦公室。
Shipping from London to Taiwan takes an average of two weeks.	從倫敦到台灣的運送時間約兩週。
You will receive your goods by the end of January.	您將於一月底前收到貨品。
The method of shipment was Federal Express.	貨品經由聯邦快遞寄出。
You will receive your order within three working days.	您會在三個工作天內收到貨品。

其他貨運資訊

Your shipment's tracking number is N2255410003.	您的貨品追蹤號碼為 N2255410003。
We have accordingly shipped your order by air freight.	我們已根據您的指示，使用空運寄送商品。
As you requested, we have wrapped the fragile items in bubble wrap.	我們已依據您的指示，用氣泡紙包裝易碎物品。
We have sent you shipping documents by air.	我們已用空運寄出貨運資料。
We have faxed you the relevant shipping information.	我們已傳真相關運送資訊給您了。

信件結語

We assure you to receive your goods on the expected time.	我們確保您會在預估時間內收到貨品。
We believe your consignment will arrive safely.	我們相信貨品會安全抵達。
Please let me know when you receive the goods.	收到貨品後請告知我們。
Please inform us when they arrived.	貨品送達後，請告知我方。
We would appreciate a confirmation after the goods reached port.	貨品到港後，請告知我們。
We look forward to doing further business with you.	希望之後還能繼續合作。

實力大補帖 Let's learn more!

貼在包裝外的警告

Fragile (Do not drop) 易碎品	Inflammable 易燃物
Liquid 液體貨物	Keep dry 保持乾燥
This side up 此面朝上	Keep upright 請勿倒置
No turning over 請勿傾倒	Handle with care 小心輕放
Keep from heat 遠離熱源	Use no hooks 請勿使用鉤子
Do not use cutter 請勿裁剪	Do not double stack 請勿堆疊
Recycled package 包裝可回收	Do not recycle 不可回收

7-02 已收到貨品的回覆

通知對方到貨狀況，信中應提及訂單號、收到貨品的日期、產品名稱等方便對方確認訂單的資訊，有特別需要溝通的狀況，也可以於「確認收到貨品」之後描述。

From:	Howard Hu
To:	Bella Jones
Subject:	Re: Shipment (Order No.12553)

Dear Bella:

Your shipment No.12553 has arrived at Taichung Port yesterday. Thank you for your prompt and timely shipment. Those items were urgently needed in order to deliver to our customers by the end of May.

As agreed, we will settle the bill immediately and we look forward to doing further business with you soon.

Best regards,

Howard Hu

貝拉您好：

運送號 12553 的貨品已於昨日抵達台中港。感謝您的快速寄送，我們非常需要在五月底前寄送此批貨物給客戶。

如雙方同意，我們會立即付款，希望日後還能再度與您合作。

胡霍華 敬上

This is to inform you that our order has arrived at Taichung Port this morning.	謹告知：我方訂購之貨品已於今早抵達台中港。
We are pleased to report that our goods have arrived safely at Tokyo Port.	很高興通知您：向您訂購的貨品已安全抵達東京港。
Our order was received on March 2.	我們已於三月二日收到貨品。
Those materials are needed urgently in order to continue our production line.	我們亟需這些材料以使生產線不致間斷。
We are able to meet our production schedule thanks to the punctual arrival of our order.	貨物如期抵達，使我們能趕上生產進度。
Thank you for expediting our order.	謝謝您快速處理我們的訂單。
Thank you for the shipment.	謝謝您的出貨。

實力大補帖 Let's learn more!

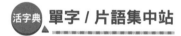

♥ 用以包裝的容器

carton 紙箱	crate 條板箱
iron box 鐵盒	wooden case 木箱
bottle 瓶子	gunny bale 麻布包
corrugated carton 瓦楞紙箱	

7-03 抱怨賣方延遲出貨

抱怨對方的出貨延遲時，要在信中寫明「本應交貨的時間」，並告知對方目前尚未收到貨品，強調希望對方能夠儘速採取行動，並請對方解釋延遲的原因。

From:	Bella Jones
To:	Mr. Hu
Subject:	Delay in Delivery

Dear Mr. Hu:

The goods we ordered were supposed to arrive on July 10. However, we have not received them as of July 16. Would you kindly explain what is going on?

The delay causes us great inconvenience since we have promised our customer early delivery. Please take this as an urgent matter.

We look forward to a prompt reply.

Best regards,

Bella Jones

胡先生您好：

我們訂購的商品預定七月十日到貨。然而，截至七月十六日，我們仍未收到貨品。您可以解釋原因嗎？

我們已向客戶保證貨物會提前運送，因此，這樣的延遲已造成我們極大的不便，請務必緊急處理此事。

期盼您儘速回覆。

貝拉・瓊斯 謹上

複製 / 貼上萬用句 Copy & Paste

信件開頭：點明尚未收到貨運

This merchandise was supposed to be delivered by the end of April.
這批貨品本應於四月底前寄送。

According to the terms, the items should be delivered no later than September 25.
根據合約條款，貨品應於九月二十五日前寄出。

We have not received any notification that you have shipped our order.
我們尚未收到任何寄出貨品的通知。

Contrary to your notice, the shipment of our order has not arrived yet.
儘管收到您的通知，但貨品卻未送達。

Unless the order is arrived within five days, we shall have to cancel the order.
如貨物五天內未寄達，我們將會取消訂單。

We have waited for one month for the shipment of these goods.
我們已等待此批貨物一個月了。

Please tell us the reason for the delay.
請告知我方延遲的原因。

Could you please look into the reason immediately for the delay?
可否請您盡快調查延遲出貨的原因？

說明：延遲造成損失

As agreed, the punctual shipment is very important because we need to deliver to our customers at the promised time.
按雙方同意，準時送達商品對我們而言非常重要，因我們已承諾客戶會按時交貨。

We made this order with you for your attractive delivery terms.
因貴公司的送貨條件吸引人，我們才予以下單。

We must supply those goods to our dealers by this Friday.
我們必須在本週五前出貨給經銷商。

We cannot delay supplying products to our dealers.
我們無法延遲對經銷商的出貨。

As we are running out the raw materials, the items we ordered are urgently needed.	由於原物料已經耗盡，我們亟需這批貨物。	
If you haven't shipped them yet, we hope you can expedite the shipment by air.	如果貨物尚未寄出，我們希望您以急件空運寄送。	
The delay will affect our reputation and sales.	延遲出貨會影響我們的聲譽以及銷售。	
The delay will cause us losing some customers.	延遲出貨會造成我方流失客戶。	

信件結語：貨運的要求

As the goods have not been delivered, we must ask you to ship them without any delay.	貨物如尚未寄出，請您儘速寄出，勿再有延遲。
We are awaiting your reply and please specify a new delivery date.	請回覆我方，並確定新的寄送時間。
Please inform us by return about the time you can ship them with certainty.	請回覆本信，告知您確定能出貨的時間。
Please do everything possible to ensure punctual shipment.	請盡力確保準時出貨。
Please inform us the time of delivery of each shipment.	每次出貨都請告知出貨時間。
Please respond to this issue as soon as possible.	請儘速回覆此問題。

有關延遲貨運的問題，應該要一收到對方的通知就立刻調查原因，並在釐清原因後迅速回覆，以表誠意。回信時，詳細說明延遲的原因，誠懇地致歉，並提供解決方式，以一次性地解決延遲的問題。

From:	Howard Hu
To:	Bella Jones
Subject:	Apology for Late Delivery

Dear Bella:

We **apologize**[1] **deeply**[2] for the **delay**[3] in shipment. Due to **peak**[4] season, it is very difficult for us to process your order in time.

However, we have **packed**[5] your goods and are ready to ship them this **afternoon**[6]. We assure that you will receive them this Friday.

We sincerely apologize for the inconvenience. Thank you very much for your consideration.

Best regards,

Howard Hu

貝拉您好：

我們對於延遲出貨深感抱歉。目前正值業務巔峰時期，因此沒能及時替您處理完訂單。

然而，我們已替您包裝好貨品，並準備於今天下午寄出。我們確保您會在本週五收到。

造成您的不便，我們深感抱歉，並感謝您的諒解。

胡霍華 敬上

E-mail 關鍵字一眼就通

1 apologize 動 道歉　　**2** deeply 副 深刻地　　**3** delay 名 延遲

4 peak 形 高峰的　　**5** pack 動 包裝　　**6** afternoon 名 下午

複製 / 貼上萬用句 Copy & Paste

信件開頭：致歉

We appreciate you bringing up this matter.	感謝您提醒我方此事。
We would like to apologize for the delay in shipping your merchandise.	造成您的貨品延遲運送，我們深感抱歉。
We are sorry for this unfortunate delay of shipment.	我們對於延遲出貨感到抱歉。
We are very sorry for delaying shipping your goods.	對於延遲運送您的貨品一事，我們深感抱歉。
We will find out the reason and report to you as soon as possible.	我方會追查原因，並盡快回覆您。

告知已出貨

Your order has now been processed and shipped.	我們已處理好您的商品，並已出貨。
We are sure you will receive the goods tomorrow.	我們確定您明天就會收到商品。

貨運有狀況

Due to the typhoon warning, we are afraid the shipment will be delayed.	由於颱風警報，我們恐怕得延遲出貨。

Due to the bad weather, we cannot deliver the items to you.	由於天候不佳，我們無法寄送物品。
As the traffic is slow, the driver told us that he cannot deliver the goods to you on time.	貨運司機告知：由於交通堵塞，貨品無法準時送達。
Due to a great shortage of shipping space, we are afraid we cannot ship the goods on time.	由於貨運的空間不足，我們恐怕無法準時出貨。
We found our shipping department made some mistakes and has delayed several shipments, including yours.	我們發現運送部門出了差錯，因此造成您與其他顧客的出貨延遲。

商品短缺

Due to the occupied production line, it is impossible for us to produce the products in time.	由於生產線繁忙，我們無法及時生產貨品。
Due to the sudden overwhelming orders, our present inventory has been depleted.	突如其來的大量訂單造成我們的存貨告罄。
Due to high demand of this model, we are out of stock now.	由於此商品熱銷，因此目前已無庫存。
Due to a shortage of raw materials, we cannot fulfill your order now.	由於原物料缺乏，我們現在無法處理您的訂單。
Our production has been delayed due to a sudden blackout.	由於突然停電，導致我們的生產線停擺。
Due to a serious machine breakdown, the factory is closed currently.	由於機器故障的情形很嚴重，工廠目前關閉。
Due to an unexpected fire at our plant, our materials have been destroyed.	由於工廠發生火災，我們的原料全毀。

Due to the strike of the shipping company, the goods are still on board.	由於船運公司的員工罷工，貨品目前還在船上。	Part 1 入門篇
The political situation in the Middle East has forced our suppliers to go out of business.	中東地區的政治因素迫使我方的供應商結束營業。	Part 2 求職篇
As you may have heard, there was a tsunami in Indonesia which made it difficult for us to ship your order on time.	如您所知，印尼剛經歷海嘯，因此我們無法準時出貨。	Part 3 客戶往來篇

保證會盡快出貨

We will deliver your goods as soon as possible.	我們會盡快出貨。
We will ship the products within one week without delay.	我們保證會於一週內出貨。
We have arranged another shipping company to deliver your order immediately.	我們已另外安排其他貨運公司，馬上替您送貨。
We are trying to remedy this situation as quickly as possible.	我們正設法解決此問題。
We will make our efforts to deliver the goods by the first direct vessel tomorrow.	我們會盡力將貨品裝上明天的第一班直達船。
We will try our best to catch the next vessel.	我們會盡力將貨品裝上最近的船班。

信件結語：致歉 & 保證

Please accept our deepest apology for the inconvenience.	對此不便，請接受我們最誠摯的道歉。

Please accept our wholehearted apology for the late delivery.	請接受我們對此延遲出貨誠摯的道歉。
Please accept our sincere apology for the carelessness.	請接受我們對此疏失誠摯的道歉。
We hope you will forgive our mistake.	尚祈原諒我方疏失。
I assure you that our company has a solid record of on-time deliveries.	我保證本公司一向準時交貨。
I hope the delay does not cause you too much inconvenience.	希望此次延遲發貨沒有造成您太大的不便。
We assure you that such delay won't happen again.	我們保證不會再發生這種延遲的狀況。
Due to the delay, please amend the shipping date and expiry date on L/C.	由於出貨延遲，請更改信用狀上的寄送日期與截止期限。
We look forward to receiving your agreement to this new delivery date.	我們希望您能同意新的交貨日期。
We look forward to having opportunity to serve you again.	我們希望能有再次為您服務的機會。
Your understanding will be highly obliged.	非常感激您能諒解。
We will keep you informed about the further confirmation about the shipment.	我們會隨時告知您貨運的消息。

7-05 抱怨商品損壞

在「抱怨商品內容」的信件中，應清楚說明商品損壞的內容，或是拍照夾帶於電子郵件中，讓對方確認。抱怨的態度要誠懇而堅定，並於信中告知希望得到的補償方式。

From:	Mark Tsai
To:	Mr. Tang
Subject:	Damaged Goods (Order No.1553)

Dear Mr. Tang:

We have received our order this morning. However, after we opened the cases, we found several glasses broken even though they are properly **wrapped**[1]. Attached you will find the photos of the **damaged**[2] goods. I believe that I **am entitled to**[3] a **replacement**[4]. It would be great if you can send me some new ones for replacement as soon as possible. Pleasc also advise should you have other **solutions**[5].

I would appreciate your immediate attention to this matter.

Thank you.

Best regards,

Mark Tsai

親愛的唐先生：

我們已於今早收到貨品。然而，當我們打開箱子時，發現儘管商品已包裝妥當，但有些玻璃杯還是碎了。實際損傷情況請參考附檔中的照片。我相信我方有要求更換商品的權利，請盡快寄送新品以替換損壞的貨品。若您有其他處理方式，也請讓我知道。

感謝您盡速處理此事宜。

蔡馬克 謹上

1 wrap 動 包；裹　　2 damage 動 毀壞　　3 be entitled to 片 有資格

4 replacement 名 更換　　5 solution 名 解決辦法

複製 / 貼上萬用句 Copy & Paste

抱怨商品損壞

I really appreciate you deliver these products on time.	我們十分感謝您準時交貨。
Some goods are damaged due to improper packing.	有些貨品因為包裝不當而損壞。
This seems to be the result of careless handling.	似乎是因為處理不當所致。
As attached photos, you can see some goods are badly damaged when arrival.	如附件照片所示，很多商品到貨時已經損壞。
Upon examination, we have found some of the machines are severely damaged.	經檢查，我們發現有些機器受到嚴重損傷。
It seems that the goods arrived in damaged condition.	有些商品在送達時似乎就已是損壞的狀態了。
The container seemed to be stuck by a heavy object.	該箱貨物似乎曾遭重物撞擊。
We found that about 20% of the package is broken.	我們發現近百分之二十的貨品包裝破損。
The container appeared to be in perfect condition.	裝貨箱看起來完好。
We accepted to sign without questions.	我們毫不懷疑就簽收了。

We can only assume that the damage was caused by careless handling.	我們推測是由於處理不慎而導致損壞。
We would like to tell you that three desk lamps were found to be damaged when arrival.	我們必須通知您：三盞桌燈在到貨時已損壞。
When the cases were opened, some of the goods were soaked by rain.	開箱時，有些商品已被雨水浸濕。
The damaged goods appear to have been mainly caused by the faulty packing.	很顯然，包裝不慎是造成損壞的主因。
We have requested you to pack the items individually.	我們已要求您個別包裝貨品。

抱怨瑕疵品

Please be informed that there are serious defects in the goods we received.	請知悉：我們收到的貨品中有嚴重損傷的瑕疵品。
Regrettably, the items delivered yesterday were not perfect.	商品已於昨日送達，但遺憾的是，其狀態並不完好。
We are sorry that we must draw your attention to the defective goods.	很抱歉，我們必須向您指出瑕疵品的問題。
We regret that there are serious defects in the goods we received.	很遺憾，我們收到的商品有嚴重瑕疵。
The goods you sent to us are totally unacceptable.	您寄來的貨品完全不符合我方的要求。
Please let us know what we should do with those defective goods.	請告知我們該如何處理這批瑕疵品。
Your defective goods will affect our promotion next week.	您的瑕疵商品會影響我們下週的促銷活動。

要求退貨 & 換貨

These damaged goods are unsalable.	這些受損商品根本無法出售。
These goods are unusable.	這些產品無法使用。
Due to such situation, we have lost some of our important customers.	由於此狀況，我們已損失部分重要客戶。
Please inspect your factory to clarify the cause.	請檢驗工廠以查明原因。
Please send us replaced items by air as soon as possible.	請盡快用空運寄送替換商品。
Please make sure that you use cushions around the products.	請在產品周圍加上襯墊。
As the problem was due to your carelessness, we will not pay the freight charge.	由於是貴公司的疏失，因此我方不準備承擔運費。
We will return the damaged goods and hope you would replace them immediately.	我方將退回毀損的商品，並希望貴公司能盡快補寄替換的新品。
We have returned these items by mail.	我們已經寄回這些商品。
We expect you to take these damaged goods back and replace them immediately.	我們希望您能取回毀損的商品，並儘速置換。
We are returning these items by parcel post for immediate replacement.	我們將以包裹退還這些產品，並請您寄回新品。
We would like you to send new items to replace them.	請寄出新的物品以換貨。
Please send new products as soon as possible.	請盡快寄來替換的新產品。

Those damaged goods have been kept aside.

損壞產品已暫時存放起來了。

We will hold the goods in our warehouse.

我們會把貨物保管在倉庫裡。

We would be grateful if you could replace the damaged goods within three days.

如果您能在三天內置換商品，我們將感激不盡。

Please deliver the alternative goods before this Friday.

請在本週五前寄出替換商品。

要求賠償

I am afraid I will have to ask for a refund for this order.

我方恐怕將要求退款。

We need to reduce our selling price by 10% for those damaged goods.

我們必須以九折賣出那些瑕疵商品。

We hope you could compensate us for the loss.

希望貴公司能夠賠償我方的損失。

We would be pleased to learn that you assure some compensation for the damage.

我們樂於收到您保證賠償我方部分損失的消息。

A reasonable compensation cannot be avoided.

必須向您要求合理的賠償。

Please make us a discount of 15% on the invoice cost.

請給予我們發票金額百分之十五的折扣。

The surveyors of our insurance company will investigate the extent of the damage.

保險公司的調查員會來調查損傷程度。

We will forward our report and the claim to you.

我們會提供調查報告以及索賠內容給您。

其他解決方式

I would appreciate if you can send a technician to fix the machine.	希望您能派一名維修員來修理機器。
I would like to know if you will replace them or just refund to us.	我方想知道貴公司準備換貨還是退款。
Please let me know what I should do with these defect products.	請告知我方該如何處理這些瑕疵品。
We would be grateful if you could propose a settlement to solve the problem.	若貴公司能提供解決方案，我方將感激不盡。
We would be pleased if you will look into the matter right now, and let us know the reason for the damage.	希望您能立即調查此事，並告知原因。

信件結語

The surveyor's report is attached herewith.	附上調查報告給您。
Please pay more attention to our instructions in the future.	請貴公司今後更加注意我方的指示。
We hope you could take precautions to prevent a repetition of the damage.	我們希望您可以預防並避免重蹈覆轍。
We hope you will bear this in mind in handling our future shipment.	希望您將來處理貨品時，能將此事謹記在心。
We hope you will be careful on packing for our next order.	希望您下次包裝時能小心。

Part
1
入門篇

Part
2
求職篇

Part
3
客戶往來篇

Part
4
人際互動篇

7-06 抱怨商品有誤

From:	Travis Walker
To:	Oscar Wu
Subject:	Wrong order (Order No.190)

Dear Mr. Wu:

Thank you for the prompt shipment of the goods we ordered on July 10th (Order No. 190). We are pleased with it on the whole, but we found that the quantity we received does not correspond to the invoice. We have received 135 bags instead of 150 as stated on our order.

Moreover, upon checking the goods, we found that 5 bags had been severely damaged due to careless handling of fragile packing.

The wrong delivery has caused us great inconvenience since our customers are in urgent demand of these goods. We need your urgent help on this issue.

Please arrange to deliver the remaining products no later than next week, and also please send 5 more new bags of intact goods to replace the damaged ones.

Your prompt response to this issue will be much appreciated. Thank you.

Sincerely,

Travis Walker

吳先生您好：

謝謝您快速送出我們於七月十日購買的商品（訂單編號190號）。整體而言我們很滿意，但我們發現您寄送的數量與訂單不符：我們訂了 150 袋，卻只收到 135 袋。

另外，檢查貨況後，我們發現其中 5 袋商品因處理不當而導致損傷。

這個狀況造成我們很大的困擾，因為客戶需要馬上拿到這些商品，請您盡快處理此事。

最慢請於下週前補上不足的貨量，並額外寄送 5 袋未損壞的新品給我們。

希望您能盡快處理這個問題，謝謝。

崔佛斯・沃克 謹上

複製 / 貼上萬用句 Copy & Paste

抱怨數量有誤

Thank you for your prompt shipment of the goods we ordered.	謝謝您迅速寄來我們訂購的產品。
We have received 150 bags this afternoon, instead of 100 as stated on our order.	我們訂購的是一百包，但卻收到了一百五十包的產品。
Please be informed that the quantity of the shipped goods do not correspond to the invoice.	貴公司寄來的商品數量與出貨單不符。
We have found a shortage in quantity.	我們發現數量短缺。
Upon taking delivery of the cargo, we found there were only 20 boxes against 25 boxes.	檢查貨箱後，我們發現只有二十箱產品，而非二十五箱。
There's a quantity shortage for three boxes; there were only 10 sets against 12 sets.	有三箱的數目短缺，裡面只有十組商品，而非十二組。
Five sets of items are missing.	商品數量少了五組。
The delivered goods are missing by 50 pieces.	商品的數量少了五十件。
One box is missing in the delivered goods.	貨品中有一盒不見了。

抱怨品項錯誤

We found your goods did not meet the specifications outlined in our order.

我們發現您的產品並不符合我們訂單上所要求的規格。

The colors of the goods are dissimilar to your original sample.

產品顏色與樣品不同。

The size of the women's shoes we received is not what we ordered.

我們收到的女鞋並非我們訂購的尺寸。

Upon opening the case, we found it contained different items with our order.

開箱後，我們發現送來的貨品與訂單不符。

We found you sent us desk lamp No.225 instead of what we ordered: No.223.

我們發現你們寄送成 225 型號的桌燈，而非我們要的 223 型號。

Upon opening the container yesterday, we found there are three cartons of products that we didn't order.

開箱後，我們發現有三盒不在訂單內的貨品。

There are five cartons that are not ordered by us.

有五箱不是我們訂購的產品。

We supposed you have loaded another customer's products in our container.

我們推測您將其他顧客的商品裝入我們的貨箱中。

Five of the cartons were severely damaged due to careless handling of fragile packing.

因未處理好易碎品的包裝，有五箱產品嚴重損壞。

The goods we received do not correspond to any of the orders by us.

收到的商品完全不符合我們訂購的貨物。

表達不滿

This error put us in a difficult condition.

此狀況讓我們很困擾。

The wrong delivery has really caused us great inconvenience.

送錯商品已經造成我們極大的困擾。

I am very angry at receiving the incorrect items.	寄錯商品這件事讓我感到非常不悅。
We will not accept this shipment.	我們不會接受此批商品。
We will not accept these goods and will not pay for them.	我們不會接受這些貨品,也不會付款。

要求對方處理

Our present needs are completely fulfilled and we don't need the extra ten machines.	目前的數量已足夠,我們不需要多出來的十台機器。
There are several customers waiting for the products.	有好幾位客戶正在等候這些產品。
Under the circumstance, we are forced to place orders with other suppliers immediately.	在此狀況下,我們只好向其他供應商下單。
Please send the shortage items without fail within one week.	請在一週內補寄不足的數量給我們。
Please arrange to deliver the remaining products as soon as possible.	請盡快將不足的數量補寄給我們。
We still need the full quantity we ordered.	我們仍需要當初訂購的數量。
Please deliver the missing items by FedEx immediately.	請立即以聯邦快遞寄出不足的數量。
Attached file is the list about the missing products.	數量不足的產品請見附檔。
We would like to ask you to replace these items to correct size.	我們希望您能夠把這些商品換成正確的尺寸。

7-07 抱怨商品的品質

From:	Melody Su
To:	Mr. Tanaka
Subject:	Replacement of poor quality goods (Order No.1520)

Dear Mr. Tanaka:

We regret to point out that the order (No. 1520) dispatched by you is not up to usual standard quality. They are not as good as you promised them to be.

First of all, the operation system is not what we requested. Secondly, the panel color is different from the sample you sent. Most importantly, the machine automatically shuts down once in a while even without long-time operation. Therefore, we are left with no other alternatives but to ask you to replace our order with a better quality one.

Please let me know how you wish us to return the unsatisfactory product.

Thank you.

Best Regards,

Melody Su

Purchase Manager

田中先生您好：

很遺憾通知您：向您訂購的訂單（編號 1520）未達到平常的品質標準，產品品質並不如您保證的那麼好。

首先，機器的作業系統與我們所要求的不符。再者，機器的面板顏色與當初送的樣品不一樣。最重要的是，就算運轉的時間不長，機器還是會不時自動關機。因此，我們不得不向您要求更換品質正常的產品。

請讓我們知道該如何退回這個瑕疵商品，謝謝。

採購部經理 美樂蒂・蘇 謹上

抱怨：品質低劣

We are writing to you that we are not satisfied with your goods.	此函是為了通知您：我方不滿意貴公司的商品。
I am sorry to say that the quality is not so satisfactory.	很抱歉，貴公司的商品不怎麼令人滿意。
We are sorry to say that the merchandise has not met our satisfaction.	很遺憾，這批貨物無法令我方滿意。
We have to complain about the quality.	我們必須向您抱怨品質的問題。
This is to complain the quality of your products.	此函是為了向貴公司抱怨產品品質的問題。
Upon unpacking the cartons, we found the quality is far inferior to the approved sample.	開箱後，我們發現產品的品質遠低於您之前提供的樣品。
The quality of your goods is much inferior to the samples.	貴公司的產品品質遠低於樣品。
The products are not satisfactory.	產品無法令人滿意。
The quality of these goods is far from satisfactory.	這批商品完全無法令人滿意。
The products you shipped are much more inferior in quality to what we have expected.	您寄來的產品與我們預期的品質差距甚大。
Your goods seem to be roughly produced.	您的產品製作非常粗糙。

抱怨：來自客戶的不滿

Our customers are very disappointed with the products.	我方客戶對於購買的產品非常失望。

Our customers complain that the quality of the products is low.	本公司的客戶向我方抱怨產品品質低落。
Those products you sold us have caused numerous complaints.	您出售給我方的產品，已經引起許多客訴。
We have recently received many complaints from customers about your products.	我們最近收到很多客戶抱怨貴公司產品的事件。
Some of our customers have been complaining about the quality of your products.	有些客戶向我們抱怨貴公司產品的品質不佳。
The quality of the products is too low to meet the requirement for consumers.	產品的品質太差，無法滿足消費者的需求。
Sometimes we need to refund the purchase price.	我們有時候還得退費給顧客。
Due to the dissatisfaction of our customers, we have to refund the purchase price on many of the products.	因為客戶的抱怨，我方必須退費給他們。

談論後續的處理

You should control the quality of your products.	貴公司應該要控管品質。
We hope we can return unsold products.	我們希望能退回未售出的產品。
Regarding your inferior quality of your goods, we claim a compensation of US$1,000.	針對貴公司商品的低劣品質，我們求償一千美元。
Please let me know how you wish us to return the unsuitable products.	請告知我方該如何退還這些瑕疵品。

7-08 針對商品損壞致歉

回覆抱怨信件時，最好不要急著做辯解，應該積極地提供解決方式或替代方案，讓對方感受到願意解決問題的誠意，這才是最重要的。

From: Paul Stewart

To: Ms. Tsai

Subject: Re: Damaged Goods (Order No.1553)

Dear Ms. Tsai:

Thank you for your letter regarding the damaged goods arrived yesterday. We were very surprised to hear that some of the goods were damaged.

Replacement for those damaged goods has been **dispatched**[1] to you this morning. Thank you for **putting the damaged ones aside**[2] for us. We will have our shipping **driver**[3] bring them back next Monday.

We are now seeking the advice of a packing **consultant**[4] in order to improve our **methods**[5] of **handling**[6]. We hope the steps we are taking will make sure the safe arrival of all your orders in the future.

Best regards,

Paul Stewart

蔡女士您好：

謝謝您的來函，並指出昨日寄達的商品出現損壞的狀況，對於這一情形，我們也感到很訝異。

今早，我們已寄出替換品。謝謝您暫時替我們保管損壞的商品。我們的貨運司機將於下週一前往取回。

我們目前正在尋求包裝專家的建議，以改善我們的處理方式。希望我們採取的措施，能夠確保您未來所訂購的商品安全抵達。

保羅・史都華 敬上

key words ▲ E-mail 關鍵字一眼就通

1 dispatch 動 發送　　2 put sth. aside 片 放到一邊　　3 driver 名 司機

4 consultant 名 顧問　　5 method 名 方法　　6 handle 動 處理

複製 / 貼上萬用句 Copy & Paste

針對損壞 & 瑕疵致歉

We apologize for the damaged goods that you received this morning.	對於您收到損壞的商品，我們感到很抱歉。
Please accept our deepest apologies for the unsatisfactory delivery this morning.	有關今早令您不滿意的貨品，請接受我們誠摯的道歉。
We regret to hear that some of the goods are found broken and unsatisfactory.	我們很抱歉物品損壞。
We are sorry that our goods are damaged on the way.	我們很抱歉物品在運送途中受損。
We do apologize for damaging the products during the transportation.	我們對於運送途中損傷貨品深感抱歉。
We are surprised about the breakage during transportation.	我們很訝異商品竟然在運送途中受損。
We have followed your instructions on how to pack and mark the containers.	我們已按照貴公司的指示包裝以及標示外箱。
We apologize for the defective products.	我們對於本公司的瑕疵品感到抱歉。
We apologize that the shipment of goods you received contained defective items.	非常抱歉您收到的貨品中有瑕疵品。
We are sorry to hear that you received several defective products.	我們很抱歉您收到的商品中，有好幾個是瑕疵品。

回收瑕疵品 & 換貨

Please be informed that the alternative goods will arrive at 11 a.m. in replace of the defective goods sent yesterday.	我們已寄出新品替換昨日送出的瑕疵品，商品將於今早十一點寄達。
Please accept our replacement for the defective products, which was delivered on this morning at 10 a.m.	請接受本公司以替代品取代瑕疵品，新品已於今早十點寄出。
We will replace those defective products as soon as possible.	我們會盡快替換那些瑕疵品。
Our shipper will collect these defective goods.	我們的貨運公司會去回收這批瑕疵品。
We would like to ask you to send those defective goods to us by airmail at our expense.	我方希望您能以航空郵寄寄回該批瑕疵品，運費將由我方支付。
We will certainly exchange the goods.	我們一定會更換商品給您。

調查緣由

We are trying to discover what the problems were.	我們正試著找出問題所在。
We are thoroughly investigating the matter and will contact you once we have answers.	我們正在調查此事，一有結果便會與您聯繫。
We are working to find out why the goods were damaged during transportation.	我們目前正在調查貨品於運送過程中損壞的原因。
The shipping company is asked to visit you and to inspect the damage, so that they can arrange compensation.	我們已要求貨運公司前去拜訪您，確認商品的損壞情況，以便處理賠償金。
We presume that the damage was caused by rough handling in transit.	我們推測損壞是由於搬運不慎所致。

We found out the error was made in our Dispatch Section.

我們發現錯誤是由配送部門造成的。

We will send our technician to repair the machine.

我們會派維修人員去修理機器。

Some goods are covered in dust due to the construction on the road.

由於道路施工，造成某些貨品沾染灰塵。

The flood has caused our warehouse soak in water.

這場洪水導致我們倉庫的貨品泡水。

責任承擔：由己方負責

Please tell us the condition of the damage.

請告知我方損壞情況。

We will try our best to compensate you generously.

我方會盡力補償您。

We will replace the merchandise free of charge.

我們會免費置換商品。

We accept full responsibility for the damaged vases.

本公司願意負擔花瓶損壞的責任。

You will receive financial compensation for your loss tomorrow morning.

您明早會收到賠償金。

責任劃分：非己方造成之損失

We have checked with the Dispatch Section and they told us that the merchandise left here in perfect condition.

我們已與配送部門確認，物品送出時完好無缺。

We have contacted with the shipping company, and asked them to explain why the goods were damaged.

我們已聯絡貨運公司，並請他們解釋損壞的原因。

We will ask the shipping company to inspect the damage.	我們會請貨運公司前去檢查破損情況。
We will ask the shipping company to arrange compensation.	我們會與貨運公司交涉賠償事宜。
We do not think we should be responsible for the claim.	我方不認為應該要承擔此次索賠。
You should file your claim for the damage with the shipping company.	貴方應向貨運公司提出損壞賠償。
We have passed your claim to the insurance company.	我們已將您的要求告知保險公司。
The insurance company will compensate you for the losses.	保險公司將會賠償您的損失。

需要對方配合的事項

It will not be necessary for you to return the damaged products.	損壞商品毋需退回。
Those damaged goods may be destroyed.	這些損壞的貨品可被銷毀。
Would you please discard the damaged items?	可以請您丟棄那些損壞的產品嗎？

信件結語

We will try our best to avoid the same problem happen again in the future.	我方今後會避免同樣的問題再度發生。
We assure you that we are taking steps to ensure this kind of mistake will not happen again.	我們向您保證，我方已在採取措施，以確保此類狀況不會再發生。
We would like to offer you a 20% discount on your next order.	我們願意下次給予您八折的優惠。

Part
1
入門篇

Part
2
求職篇

Part
3
客戶往來篇

Part
4
人際互動篇

7-09 針對商品錯誤致歉

　　若是遇上寄錯商品給買方的情況，除了誠摯地道歉之外，務必要提供後續的處理方法。若買方因為此次運送的錯誤蒙受損失，最好也提供對方一定的補償。除此之外，若寄了完全不同的商品給買方，也要記得告知己方準備如何處理那些寄錯的商品。

From:	Oscar Price
To:	Travis Tung
Subject:	Reply: Wrong order (Order No.190)

Dear Mr. Tung:

Thank you for **bringing**[1] the **issue**[2] to our **attention**[3]. I am really sorry for the inconvenience **caused**[4], and this mistake is **entirely**[5] **on our own**[6].

According to your order, the quantity should be 150 bags of papers, not 135. Our Packing Department had carelessly made the mistake, and we will make sure not to make such issue happen again. Besides, for the 5 bags of damaged goods, we will also take full responsibilities and make compensation to you for the **loss**[7].

Tomorrow morning we will send you the **correct**[8] products **at our expense**[9], and you should be receiving the goods by next week.

Once again, I am really sorry for the mistakes. We promise this will not happen again, and hope to continue providing you with our usual quality service in the future.

Yours sincerely,

Oscar Price

董先生您好：

感謝您來信告知運送錯誤的消息，這完全是我方的疏失，非常抱歉造成您的困擾！

經查，您下的訂單數量應為150包紙張，而非135包，是我們的包裝部門不小心弄錯了，

我們會確保這種情形不再發生。而關於運送過程中損壞的 5 袋商品，我們也會負責並給予補償。

明天早上我們會寄出正確的商品，運費將由我方支付。預計將於下週前送達。

再一次針對此次錯誤向您致歉，我們保證未來不會再發生同樣的問題，也希望往後能持續與您合作，提供您優良的服務品質。

奧斯卡‧普萊斯 謹上

 E-mail 關鍵字一眼就通

1 bring 動 拿來；使來到　　2 issue 名 問題　　3 attention 名 注意

4 cause 動 導致；引起　　5 entirely 副 完全地　　6 on one's own 片 獨自

7 loss 名 損失；虧損　　8 correct 形 正確的　　9 at one's expense 片 某人花費

 複製 / 貼上萬用句 Copy & Paste

表達歉意

Thank you for bringing the issue to our attention.	謝謝您告知我們這個問題。
This mistake is entirely on our own.	這完全是我方的疏失。
This kind of mistake is very unusual.	這樣的錯誤很不尋常。
We apologize for our carelessness.	我們對此疏忽感到抱歉。
We apologize for any inconvenience it may cause you.	我們對造成的不便感到抱歉。
Please accept our sincerest apologies for the unfortunate mistake.	針對本次錯誤，請接受我們誠摯的道歉。

數量上的錯誤 & 後續處理

According to your order, the quantity should be 25 boxes of papers, not 20.

根據訂單，您的購買數量應為二十五箱紙，而非二十箱。

We have just checked your order, and realized we had sent you extra five boxes of toys.

我們檢查過您的訂單，發現多寄了五箱玩具給您。

Please accept our sincere apologies for delivering 10 instead of 15 laser printers to you.

關於本公司應該寄送十五台雷射印表機，卻寄成十台的這件事，我們深感抱歉。

We are sorry for our stock shortage issue.

我們為商品缺貨向您致歉。

The shortage is caused by our careless Packing Department.

貨物短缺起因於包裝部門的粗心。

We are prepared to accept any claims.

我們準備接受貴公司求償。

I have spoken to the shipping department and hope they will not make this kind of mistake again.

我已與運送部門談過，希望他們不再犯同樣的錯誤。

Would you accept that we resupply the remaining items in the next shipment?

不知您可否接受下次出貨時補足缺的貨品？

送錯產品 & 後續處理

We thank you for your mail, claiming the wrong size of the shoes.

感謝貴公司來函申訴鞋品的尺寸不合。

We are sorry to find that our driver had delivered the goods to the wrong address.

很抱歉，我們的貨運司機送錯地址了。

The mistake was caused by the oversight of our production department.

這是本公司生產部門的疏忽。

This mistake was made due to the staff shortage during the busy season.	因為正逢旺季，又遇上人手不足的狀況，所以造成這個錯誤。
The women's dresses we have sent to you were the wrong size, and we are very sorry about that.	我們非常抱歉寄了錯誤尺寸的女性洋裝給您。
Your account will be credited with the invoiced value.	寄錯的產品款項將從您的帳單扣除。
We have arranged a shipping company to collect the mis-shipped products.	我們已安排貨運公司前去取回送錯的產品。
Tomorrow morning we will send you the correct products at our expense.	我們明早會寄出正確的商品給您，運費由我方負擔。
We apologize for our mistake, and had already sent the correct items to you this morning.	對於這次的疏失，我們深感抱歉。今早已將正確的商品寄給您了。

信件結語：表現出積極的態度

We apologize again for this mistake.	再次表示歉意。
We will take necessary steps to correct our mistake.	我們會採取必要步驟來更正錯誤。
We will take this problem seriously.	我們會嚴肅看待此問題。
We will do our best to make sure it will not happen again.	我們會確保此類錯誤不再發生。
We hope to provide you with our usual quality service in the future.	本公司希望今後也能提供您高品質的服務。
We appreciate your business and value our relationship.	我們重視您的惠顧以及我們的良好關係。

From: Mr. Tanaka

To: Melody Su

Subject: Reply: Replacement of poor quality goods (Order No.1520)

Dear Melody:

Thank you for your letter informing us the problem with our product. I am really sorry for your bad experiences about the frustrating product. This problem has caused our great concern.

We have passed the defective goods to our engineers for inspection. And found that the faults have been caused by one of the problematic assembly line.

Please send us your unsold machines back, the delivery fee will be reimbursed at our expense. We will send replacement right away, and will do our best to improve our quality of service in the future.

Sincerely yours,

Tanaka Ryu

Manager

美樂蒂您好：

謝謝您來信告知產品的問題，造成您這麼差的使用體驗，真的非常抱歉，這個問題已引起我們極大的關注。

我們已將有瑕疵的產品送交工程師檢查，並發現錯誤出在其中一個有問題的生產線上。

請將未售出的產品寄回，我們會負擔所有運費。替換的商品會馬上補寄給您，我們也會致力於改善服務品質。

經理 田中龍 謹上

針對品質不良致歉

Thank you for your mail dated May 20, pointing out the faults in our goods supplied by us.	謝謝您五月二十日來信，告知我方提供的產品有瑕疵。
Thank you for pointing out the unsatisfied merchandise.	謝謝您讓我們注意到此件令人不滿的商品。
We regret your dissatisfaction with our products.	我們很遺憾您不滿意我方的商品。
This problem has caused our great concern.	本事件已引起我方的高度關注。
We agree that the products are not perfect.	我們同意產品不夠完美。
We are sorry our products do not live up to your expectations.	很抱歉，產品沒有達到您的要求。

後續處理

Please send us your unsold goods.	請寄回未售出的商品。
The cost of shipment fee will be reimbursed at our expense.	運費將由我方承擔。
We will send replacement products right away.	我們會立刻寄出替換的新品。
We are confident that the replacement will make you satisfied.	我們相信更換的產品會讓您滿意。
We will do our best to improve our quality of service.	我們會盡力提高服務品質。

Unit 8 付款及催款
About the Payment

8-01 準備付款

準備付款時，需再次確認付款的相關資訊，如金額、期限、付款方式、對方的銀行帳號等，以確保付款順利。

From:	Danny Wang
To:	Ms. Clark
Subject:	Payment Notice (Order No.991AL110)

Dear Ms. Clark:

We are looking forward to receiving the invoice for May. Could you please send that to us first? We will transfer the funds to you on due date.

Best regards,

Danny Wang

克拉克女士您好：

我們在等五月份的帳單，可以請您先寄給我們嗎？付款日到時，我們會轉帳給您。

王丹尼 謹上

複製 / 貼上萬用句 Copy & Paste

信件開頭

Thank you for mailing the bill for May. | 謝謝您寄來五月份的帳單。

Thank you for the bill for May, 2018. | 謝謝您寄來二〇一八年五月份的帳單。

Our records show that we have not received payment for bill of May. | 根據我們的紀錄，我們仍未收到五月份的帳單。

Please mail us an itemized statements and proof of payment. | 請寄發票明細給我們。

The bill was spotted and couldn't be used now, would you please send us a new one? | 我們弄髒了帳單，能請您再寄一份嗎？

We wonder if you missed out the 5% discount that we agreed on April 5? | 不知您是否遺漏了四月五號所協議的百分之五折扣呢？

討論付款條件

We cannot accept your prices quoted in British Pound. | 我們無法接受以英鎊報價。

Please quote the price in Euro. | 請以歐元報價。

Shall we discuss about payment terms? | 我們可否討論付款條件？

We would like to pay in cash. | 我們想要以現金支付。

Do you accept US dollars? | 您接受美金嗎？

We would like to pay by check. | 我們想要以支票支付。

Do you accept credit cards? | 您接受信用卡嗎？

Do you accept credit cards issued by an overseas bank? | 您接受海外銀行發行的信用卡嗎？

We would like to pay by international money orders.	我們想要以國際匯票支付。
We would like to pay initially of $1,000, and the remainder in three installments.	我們會先支付一千元，餘額分三期付款。

要求對方配合的事項

Would you please send us our invoice in PDF file?	可否請您將發票以 PDF 檔寄給我們？
Please send us monthly bills by e-mail.	請每月以電子郵件寄送帳單給我們。
The payment will be made by sending you a bank check.	我們會寄給您銀行支票支付款項。
Please fax us the name of the remitting bank and the remittance number.	請傳真告知匯款銀行與帳號。
Please inform us your bank's name and your account number.	請告知您的銀行名稱與帳號。

實力大補帖 Let's learn more!

深入學 資訊深度追蹤

📍 支付方式：匯付（Remittance）

泛指付款人透過銀行或其他途徑將款項匯給收款人。

01 信匯（Mail Transfer, M/T）：買方將款項交給在地銀行，再透過郵遞寄交賣方所在地的銀行，委託其將貨款交給賣方。

02 電匯（Telegraphic Transfer, T/T）：買方直接向在地銀行申請，通知賣方所在地的銀行直接付款給賣方，速度較快捷。

03 票匯（Demand Draft, D/D）：買方直接向在地銀行購入匯票，寄給賣方。

8-02 已付款通知

此類信件是買方已付款後，寄給賣方的通知信。撰寫此類信件時，記得要告知對方正確的訂單號碼以及金額，以便對方確認貨款。

From: Danny Wang

To: Ms. Clark

Subject: Payment Notice (Invoice No.22256)

Dear Ms. Clark:

I'm writing to inform you that we have transferred US$6,000 as payment for order No.22256. Please check.

Furthermore, we are pleased with the way you processed the order. The order arrived on time and could be put for sell without delay.

We thank you for your prompt attention on our order.

Best regards,

Danny Wang

克拉克女士您好：

我們已將訂單編號 22256 號的六千美金轉到您的戶頭，請查收。

此外，我們很滿意您處理訂單的過程。貨品準時送達，因此沒有延誤到我方的銷售計畫。

謝謝您快速地處理此筆訂單。

王丹尼 敬上

付款通知

We are writing to inform you that we are now transferring the payment to you.

謹以此函通知：我們準備要轉帳給您了。

We are pleased to send our check of $5,000 US dollars in payment of your invoice No.223.

我們很樂意寄出五千美元的支票，以支付貴公司的 223 號發票。

The bank has credited your account with the amount of the check.

銀行已把支票上的金額轉入您的帳戶。

Referring to your payment request of May 20, 2018, we have already applied for remittance to our bank.

根據您二〇一八年五月二十日要求之付款事宜，我們已向我方銀行申請匯款。

We have made the remittance today to your account with The Taipei Branch of The China Trust Bank.

我們今天已經透過中國信託銀行的台北分行匯款給您。

The remittance number is No.55362.

匯款號碼為 55362 號。

We have asked our bank to issue a check for the invoice No.251.

我們已要求我們的銀行準備支票，以支付 251 號發票。

We are pleased to send a postal order for $5,000.

我們很樂意寄出面額五千元的郵政匯票。

We have sent you a bank draft for $2,000 in settlement of invoice No.6553.

我們已經寄給您一張面額為兩千元的銀行匯票，用以支付 6553 號發票。

The amount has been transferred to your account.

金額已經轉至您的戶頭。

The amount of $2,000 US dollars will be paid to your bank account.

兩千元美金將會匯進您的銀行帳戶。

We have remitted $1,500 to your bank account.	我們已匯一千五百元至您的銀行帳戶。
We have made the first payment on time.	我們已準時繳付第一筆款項。
We will prepare the second payment soon.	我們近日會準備好第二筆款項。
I will send a check of $1,500 for the final payment.	我會寄出一千五百元支票，以支付最後剩餘的款項。
We have instructed the bank to transfer $3,000 to your account.	我們已指示銀行匯三千元至您的帳戶。

其他細節說明

We included the details of the name of the remitting bank and the remittance number.	我們已附上匯款銀行以及匯款號碼等細節。
The postal check can be cashed at any post offices.	這張郵政支票可在任何郵局兌現。
Please confirm if you have received the check.	請確認您是否收到支票。
We look forward to your confirmation on receiving the payment.	我們希望收到您確認收到款項的訊息。
When you receive the remittance, please inform us.	當您收到匯款後，請告知我們。
Please contact us if you haven't received any notifications on our remittance.	如您尚未收到匯款通知，請與我們聯絡。
We also faxed a copy of remittance slip to you.	我們已傳真匯款成功的收據給您。
We will fax you the Application for Remittance.	我們會將匯款證明傳真給您。

信件結語：感謝賣方的配合

The entire pack of goods arrived in good condition.	貨物安全抵達。
The goods arrived successfully on May 21.	貨物在五月二十一日安全抵達。
Thank you for arranging our order promptly.	謝謝您迅速處理我方的訂單。
Thank you for shipping our order promptly.	謝謝您迅速寄出貨物。
We appreciate the efficiency of your work.	我們感激貴公司的高效率。
Thank you again for your prompt delivery.	謝謝您的快速出貨。
We hope to do business with you again very soon.	希望近日能再與您交易。

實力大補帖 Let's learn more!

深入學 資訊深度追蹤

📍 與「花費」有關的單字

01 pay 表示「付給某人的工酬」，等同於 salary, income, wages。

02 payment 指「付款的過程」或「收取的款項」。（分期付款也可以用 payment 表示，如 pay one's bill in monthly payments. 以每月扣款的方式付帳。）

03 cost 是獲得某物所需的付出，可以是「特定金額」或「抽象的代價」。

04 expense 的意思與 cost 相近，但更常用於「經常性的開支」或「固定支付的費用」，如租金、薪資、水電等。

8-03 收到款項

　　賣方在收到款項後，同樣也需回覆有關訂單號碼以及收到的款項資訊，與買方核對是否有誤；除此之外，還可以通知買方貨品的最新狀況，讓買方更清楚目前的作業流程，並加深信賴感，以利日後的合作。

From:	Frank Smith
To:	Nancy Liu
Subject:	Payment Confirmation (Order No.20235)

Dear Nancy:

The bank has informed us this afternoon that we received $5,000 for your order No.20235. Thank you for the payment.

The goods you ordered will be shipped from Keelung Port tomorrow. I will keep you updated about the shipment schedule.

As always, thank you for doing business with us.

Best regards,

Frank Smith

南西小姐您好：

銀行通知我們，今天下午已收到您訂單號 20235 的匯款，謝謝您的付款。

您訂購的貨品明日將從基隆港出貨，我會持續通知您有關貨運的消息。

感謝您一直以來的合作。

法蘭克・史密斯 敬上

複製 / 貼上萬用句 Copy & Paste

通知對方已收到款項

This mail is the confirmation of receipt of all fees.	此郵件是為了通知您：我們已收到所有的費用。
We have confirmed the receipt of your remittance.	我們確認收到您的款項。
Your payment has been received.	已收到您的款項。
Please be informed that full payment has been received.	茲告知已收到所有款項。
Thank you for your mail this morning about your remittance.	謝謝您今早通知我方已匯款的消息。
The bank informed us that your remittance for your order was received on November 11.	銀行通知我們，已於十一月十一日收到您的匯款。
We hereby confirm the receipt of $5,000 in our account this morning.	在此確認：我們已於今天早上收到五千元的款項。
We have received your remittance of 5,000 US dollars.	我們已經收到您五千美元的匯款。
Thank you for sending your remittance for invoice No.567.	謝謝您匯來 567 號發票的款項。
It was good to receive your check for $5,000 this morning.	很開心今早收到您五千元的支票。
We are pleased to receive your check of $2,000.	我們很高興收到您的兩千元支票。
Your check No.6652314-1 in payment of invoice No.2523 was received this morning.	您用於支付 2523 號發票的支票（編號 6652314-1）已於今早送達。

Thank you for your postal order for $10,000 in payment of our invoice No.223.

謝謝您寄來金額一萬元的郵政匯票，用以支付 223 號發票。

進一步處理

Please advise us which bank you remitted.

請告知我方您的匯款銀行。

We will address all invoices to your accountant later.

我們之後會將發票寄給您的會計。

A receipt will be sent to you.

發票會寄送給您。

信件結語

All fees have now been paid in full.

所有款項均已結清。

The money has been credited to your account.

該筆金額已抵付您的帳款。

Thank you very much for your payment.

謝謝您的付款。

We are pleased that the procedure of processing this order was done properly.

我們很開心此次交易順利。

We look forward to serving you again.

希望能夠再為您服務。

We look forward to having the opportunity to do business with you again soon.

希望近日能夠再與您合作。

We hope to receive other orders from you again in the near future.

希望不久之後能再收到您的訂單。

8-04 請求延遲付款

　　商業往來中，付款行為是極為重要的一環，除了涉及這一批貨物之外，還會影響公司的聲譽，像是本章第六單元所提到的「信用調查」，就會以付款行為作為主要的判斷依據。

　　因此，需要請對方諒解的「請求延遲付款」信件，一定要好好說明本身所遇到的困難以及突發狀況，表示自己絕非刻意欠款，以誠懇的態度請求對方諒解以及延長付款期限。

From:	Minnie Chen
To:	Ms. Nagazawa
Subject:	Extension of Payment (Invoice No.211)

Dear Ms. Nagazawa:

Referring to[1] your invoice No.211 dated April 30, due for payment on May 20, we have to inform you that we are unable to pay you the **sum**[2] of $10,000 at this moment.

Due to the economic **downturn**[3], customers who usually buy our goods have been very **hesitant**[4] in making purchase recently.

Despite[5] our desperate efforts in pushing for payment, cash is coming slowly. Therefore, we would ask for a two-month **extension**[6] for the payment.

We do hope you will agree to the extension of the deadline. Look forward to your confirmation.

Thank you!

Best regards,

Minnie Chen

中澤小姐您好：

關於您四月三十日的所開的 211 號發票，其付款日為四月三十日，但我們必須通知您，本公司目前尚無法支付一萬元款項。

由於經濟蕭條，我們的常客最近在購買產品時變得很猶豫。

雖然本公司極力籌措，但進帳速度仍緩慢。因此，希望您能同意延後兩個月付款。

希望您能同意此提議，並盼望得到您的確認，謝謝！

陳米妮 敬上

 E-mail 關鍵字一眼就通

1 refer to 片 提到　　2 sum 名 總數　　3 downturn 名 經濟衰退

4 hesitant 形 猶豫的　　5 despite 介 儘管　　6 extension 名 延期

 複製 / 貼上萬用句 Copy & Paste

致歉的說法

We are sorry that we were unable to settle our account on time.	很抱歉我們無法準時付款。
We are sorry we are unable to pay your draft at maturity.	很抱歉，我們無法在到期日之前付款。
We are very sorry for the delay in the payment for the bill dated May 22.	很抱歉，尚未支付您五月二十二日的帳款。
Please accept our apologies on late payment.	請原諒本公司延遲付款。
We are sorry for not meeting the payment deadline.	很抱歉沒有準時付款給您。
We are sorry for taking so much time to pay.	我們很抱歉這麼晚才付款。

We sincerely apologize for any inconvenience caused by the delay in payment.	我們對於延遲付款造成的不便深感抱歉。	Part **1** 入門篇

解釋延遲的原因

We would like to explain to you our awful situation.	我們需要解釋一下我們目前的慘況。
A sudden fire occurred at our factory, causing us great losses.	一場意外的工廠大火讓我們損失慘重。
A large number of our goods were destroyed.	我們有大量貨品遭摧毀。
Due to the flood in our factory, we are still waiting for the compensation from the insurance company.	我們工廠受洪水波及，目前仍在等待保險理賠。
The economics is declining recently.	近來經濟十分蕭條。
Our market situation is very terrible.	我們的市場狀況很差。
Our customers have not paid us yet.	我們的客戶尚未付款。
Recently, there are sharp declines in sales in our markets.	近來我們的生意驟減。
Despite our best efforts in collecting funds, cash is coming slowly.	儘管我們盡力籌款，進帳的速度仍相當緩慢。
Due to the dropped sales recently, we have financial difficulties.	因為近來業績下滑，使我們財政吃緊。
We are now in a serious financial crisis.	我們目前有嚴重的財務危機。
Due to the devaluation of dollars, there is a setback in business.	美元的貶值導致我們的市場衰退。

提出延期的要求

Would it be allowed if we pay a little later than usual?	可否容許我們比平常晚一點付款呢？
Would you please allow us to postpone the settlement of the account?	能請您同意我們延後付款嗎？
Would it be possible if we confirm the payment by July 20?	可否容許我方在七月二十日前付款呢？
Would it be possible for us to delay the payment for 20 days for this month?	是否可以容許我們延遲二十天支付本月帳款？
Could you allow us to pay you $1,500 by this week, and the remainder to be paid next week?	您能否同意我們本週先付一千五百元，下週再付餘款？
Would you invoice us until next month?	可否下個月再開發票？
Would it be possible for us to make a late payment for the bills of last month and this month?	上個月和這個月的帳款可否延遲支付呢？
Please agree for the delay payment for 20 days.	請同意我們延遲二十天付款。
Please accept our request for an extension of the deadline for payment.	請接受我們延長付款期限的請求。
Please extend one more week for the payment.	請再給予一週付款的寬限期。
We would like to ask you to reschedule the payment deadline to December 31.	我們希望您能重新安排付款日期至十二月三十一日。
We would appreciate if you give us another fortnight to settle the payment.	如果貴公司能夠再寬限兩週，我方將感激不盡。
We wish to ask you if it is possible to pay in another 20 days.	我們希望您能夠再寬限二十天。

We would like to ask you for another 20-day payment extension.	希望您能同意延後二十天付款。	
We would like to make a proposal for extending the payment.	我們希望能延後付款。	
We would like to ask for your permission for payment in three installments.	希望您能同意我們分三期支付。	
We will send you one third of the payment, which is $3,000.	我們會寄出三分之一的款項：三千元。	
We will send another third on June 20, and the balance on July 10.	我們將在六月二十日寄出另外三分之一，並於七月十日結清餘款。	
We will try to clear the balance within the next three weeks.	我們會盡力在三週內結清餘款。	

信件結語

We are sorry that we are forced to make this request.	我們很抱歉，迫不得已才向您提出此要求。
As you are aware of, we have always settled our accounts promptly.	如您所知，我們一直都是準時付款。
We hope you could understand our situation, and wait until the matter is solved.	希望您能寬容我們的情況，直到事情解決。
We hope you can understand our present situation.	希望您能體諒我們的現況。
We may default on the contract if the payment cannot be postponed.	如果無法延後付款，我們可能就必須違約。
We look forward to receiving your favorable reply.	我方希望可以得到您善意的回應。

8-05 接受延遲付款

若付款方發生了什麼難以解決的困難，在情況允許的前提下，藉由答應延遲付款來釋出善意，對未來雙方的合作關係很有助益。不過，也必須考量公司自身的情況，來判斷是否能答應對方延遲付款的條件。

From:	Alex Chou
To:	Mr. Hu
Subject:	Re: Extension of Payment (Order No.9912)

Dear Mr. Hu:

Thank you for writing to us **frankly**[1] about your **inability**[2] to settle your account. We understand it's an **unfortunate**[3] **exception**[4] due to the **earthquake**[5]. Therefore, we are willing to **grant**[6] you the extension you asked for.

Best regards,

Alex

胡先生您好：

謝謝您來信，據實以告有關您無法付清款項的事。我們能理解發生地震是不幸的意外。因此，我們願意接受您的延期付款要求。

艾力克斯 謹上

 E-mail 關鍵字一眼就通

1 frankly 副 率直地 　　**2** inability 名 無能力 　　**3** unfortunate 形 不幸的

4 exception 名 例外 　　**5** earthquake 名 地震 　　**6** grant 動 同意

複製 / 貼上萬用句 Copy & Paste

信件開頭

Thank you for your mail explaining the reason why you cannot settle the account on time.

謝謝您來信解釋不能準時付款的原因。

Thank you for your mail concerning the outstanding balance on your account.

謝謝您來信提及未結清餘額的相關事項。

We are obliged to hear from you regarding your overdue payment.

我們感激您提供有關過期帳款的消息。

We have received your mail requesting the postponement of payment.

我們已收到您要求延期付款的郵件。

答應延遲付款

We can fully understand your financial conditions.

我們完全可以理解您的財務狀況。

We are aware of your financial difficulties.

我們了解您所面臨的財務困難。

We can understand the situations which might lead to late payment.

我們能夠體諒某些導致延遲付款的情況。

We can accept your explanation about your market situation.

我們可以接受您市場狀況不佳的解釋。

We are sorry about the earthquake happened in your country, and are willing to allow you to extend the payment for another six weeks.

我們同情貴國地震後的處境，所以同意您延後六週付款。

We would like to do our best, and allow you to pay the balance before Nov. 20.

我們很願意幫忙，因此同意您於十一月二十日前付清款項。

🖰	As you have always settled the accounts on time, we will allow you a postponement of payment.	以往您都如期繳款，因此我們打算同意您此次延期付款的要求。
🖰	Thanks to the regularity of your payments; therefore, we reply favorably to your request.	感謝貴公司以往總是準時付款，因此我們願意答應您的要求。
🖰	As we have done business with each other for a long time, we will defer payment for another five working days.	由於我們已經合作了好一段時間，因此我方願意寬限您五個工作天。
🖰	Thanks to our good business relations, we would agree to your proposal to postpone the payment.	由於我們以往良好的合作關係，我們願意讓您延期付款。
🖰	As you are a valued customer of ours, we would agree to extend your credit period by one more week.	由於您是我們寶貴的客戶，我們樂意將信用期間往後延長一個星期。

有條件地答應

🖰	We allow you to extend your payment, but payment must be made by the end of July.	我們同意您延期付款，但堅持必須在七月底前繳清。
🖰	We have no choice but to accept your proposal of extending the payment.	我們沒有其他選擇，只能同意您延後付款的提議。
🖰	We found it difficult to agree to your proposal, though.	雖然我們覺得很難同意您的提議。
🖰	Nevertheless, we will still agree with your request.	但是，我們還是答應您的請求。
🖰	Accordingly, please settle the account no later than December 31.	因此，請勿超過十二月三十一日付款。
🖰	We agree to the extension, but only until December 15.	我們同意延期，但只能延至十二月十五日。

We suggest that you make a partial payment of $1,500 now, and pay the balance by the end of August.

我們建議您先付一千五百元給我們，再於八月底前付清餘額。

信件結語

We sincerely hope that your financial problems will be solved soon.

我們誠摯地希望您的財務問題能盡快獲得解決。

We hope that your present problem will be solved soon, and hope you will find a rapid solution.

我們希望您的困境只是暫時的，盼您可以盡快找到解決辦法。

With enough time, we believe you will clean off the balance soon.

我們相信您能在足夠的時間內付清款項。

It will cause us some financial embarrassment if you delay the remittance again.

如果您再次延遲，將會導致我們財務困難。

After we receive your remittance, we will e-mail you the confirmation.

待收到您的匯款，我方將以電子郵件通知。

實力大補帖 Let's learn more!

活字典 單字 / 片語集中站

📍 無力結清款項

behind schedule 比預定時間晚	defer 推遲；延期
in debt 負債	outstanding 未償付的
overdue 過期的	postponement 延期；延緩
settlement 清算；結帳	transient 短暫的

拒絕延遲付款

在「拒絕延遲付款」的信中，要明言告知自己公司也有困難，因此無法答應對方的延遲付款要求，但可提供替代方案，讓對方有所選擇。

From:	Kevin Johnson
To:	Andy Lu
Subject:	Re: Extension of Payment (Order No.991)

Dear Mr. Lu:

Regarding your proposal in the last mail for extending the payment **deadline**[1], we are really sorry about the losses and inconvenience caused by the **flood**[2] in your **country**[3].

However, please understand that we have our financial difficulties as well. Therefore, we are willing to accept a **partial**[4] payment of 50% of the amount by May 20, with the **remaining**[5] 50% being paid by June 20. We think that this **alternate**[6] plan may be a good solution.

I hope this plan meets with your request. If you have any questions, please feel free to contact me.

Best regards,

Kevin Johnson

盧先生您好：

您上次於來信中提到，希望能延長付款期限，關於這一點，我們非常同情貴國遭遇洪水災害，導致您蒙受損失，也帶來不便。

但希望您體諒本公司也有財務吃緊的狀況，因此，我們能接受的是，於五月二十日前付一半的款項，剩下的一半則於六月二十日前付清。我們覺得此替代方案會是很好的解決方式。

希望您能同意。如有任何問題，請隨時聯絡我。

凱文‧強森 謹上

Part
1
入門篇

Part
2
求職篇

Part
3
客戶往來篇

Part
4
人際互動篇

key words ▲ E-mail 關鍵字一眼就通

1 deadline 名 最後期限　　**2** flood 名 洪水　　**3** country 名 國家

4 partial 形 部分的　　**5** remaining 形 剩下的　　**6** alternate 形 供替換的

複製 / 貼上萬用句 Copy & Paste

拒絕的說法

We have considered the proposal you mentioned in your mail.	我們考慮過您於信中的提議。
We regret that it's hard for us to accept your proposal.	很抱歉,我方難以接受您的提案。
We cannot accept your request.	我方無法接受您的請求。
We of course understand your difficulties, but we have ours, too.	我們當然了解您的困境,但我方也有困難。
Your failure to pay on time has caused great inconveniences to us.	您的逾時付款已造成我方極大的不便。

拒絕的理由

The offer we gave you is on a small profit basis.	我方給您的報價已經很低,只給我們賺一點點而已。
However, we need to pay our suppliers as well.	然而,我方也需要付款給供應商。
You have delayed your payment frequently.	您時常延遲付款。
As our margin is very little, we cannot agree to the extension of the payment.	由於我們的利潤很少,所以無法同意延期付款。

信件結語：請對方付款

Therefore, we still need to ask you to settle the account at once.	因此，還是要請您立即付款。
Otherwise, we are compelled to take legal actions.	否則，我們就必須採取法律行動。
If you failed in clearing the account on time, we will claim damages for your failure to honor the contract.	如果您無法準時付清款項，我方就會依約要求賠償。
We must ask you to arrange payment immediately.	我方必須要求您立即付款。
We must insist on payment within ten days.	我方必須要求您於十天內付款。
We therefore propose to give you ten days to clear your account.	我方請您於這十天內結清帳款。
We must ask you to settle this account within the next few days.	請你務必於這幾日內結清這筆帳款。
We hope that you clear the balance without delay.	希望您能準時結清餘款。
An early remittance will be highly appreciated.	敬請提早匯款。
If the payment is not made in accordance with the contract, I am afraid we will cancel the order.	如果貨款沒有根據合約付清，我方恐怕要取消這筆訂單。

8-07 第一次催款信

第一次的催款信，態度須有禮，並簡單表明催款原因，口氣切勿過於責難，以免破壞往後的合作關係。

From:	Terry Chang
To:	Mr. Miller
Subject:	Payment Reminder (Order No.991AL110)

Dear Mr. Miller:

We would like to remind you that our April statement amounting to $5,000 is overdue. We would be grateful if you could send us your check as soon as possible.

Best Regards,

Terry Chang

米勒先生您好：

我們想要提醒您四月份結算金額為五千元，付款期已過。若您能盡快將支票寄出，我們將感激不盡。

張泰瑞 謹上

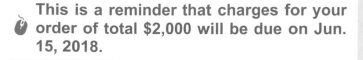

複製／貼上萬用句 Copy & Paste

告知尚未收到款項

 This is a reminder that charges for your order of total $2,000 will be due on Jun. 15, 2018.

提醒您：訂單款項兩千元，將於二〇一八年六月十五日到期。

Please kindly note that our statement of account dated May 10 is still waiting for settlement.	謹此通知您五月十日的帳款仍未付清。
This is to inform you that payment on the invoice No.52581 is now more than 60 days overdue.	通知您：52581 號發票的貨款已逾期六十日。
We have to remind you that the payment for your last order is due three weeks from the date of invoice.	我們必須提醒您：您的貨款已經延遲三週了。
I am writing to remind you that we have not received your payment for the order which was delivered to you last week.	此信是想通知您：貨品在上週已經寄給您，但我方尚未收到您的貨款。
This mail is to remind you that your remittance for last order has not been made yet.	我們想通知您：上次的貨款尚未付清。
Our records show that your account has not been paid yet.	我們的紀錄顯示：您的貨款尚未繳清。
The bill of $3,500 is overdue.	您有一筆三千五百元的帳單已超過支付期限。
Until now, we have not received your payment.	我方到現在仍未收到您的款項。
Up to this date, it appears that we haven't received your remittance.	我方到現在仍未收到您的匯款。
We wonder why the payment is still not paid.	帳款仍未付清這件事讓我們感到奇怪。
We expect the receipt of the payment two weeks ago, but we still have not received your check until now.	我方原本預期兩週前收到帳款，但現在仍未收到您的支票。
As you know, the due date for payment was March 28, but we still haven't received your remittance.	如您所知，付款到期日為三月二十八日，但我方仍未收到匯款。

We are very concerned as you have always been prompt in making your payments.

由於您總是準時付款，因此我方十分關切這次的情況。

As you are usually very prompt in setting your account, we would like to know why we have not received the payment.

由於您一向不拖欠付款，因此，我方想要了解為何至今仍未收到款項。

較緩和的說法

I believe the seasonal rush must keep you very busy and the invoice was overlooked.

相信業務高峰期肯定讓您非常忙碌，才因此看漏了帳單。

Just in case you have misplaced your copy of invoice, we will send you again.

為防您找不到那份帳單，我們會再寄一份給您。

In case you may not receive the statement of account, I have attached the files showing a balance of $1,500 unpaid.

為防您沒收到我們的對帳單，我再一次附上檔案，欠款餘額為一千五百元。

Perhaps you are unaware that invoice No.552 is still outstanding.

或許您沒有注意到 552 號帳單尚未付款。

It might be an oversight.

也許您只是漏看了。

We understand that delays can happen for various reasons.

我們了解可能因為各種原因造成延遲付款。

We understand sometimes payments are overlooked.

我們可以理解有時候會漏看帳單。

We are wondering if the invoice has been overlooked.

我們在想您是否漏看了帳單。

We wonder if you have overlooked the invoice No.6553 for $5,000, which was due two weeks ago.

不知您是否漏看了金額為五千元的 6553 號發票？這筆款項已逾期兩週。

We understand that our invoice may not have reached you yet.	我們了解您也許尚未收到發票。
We would like to take this opportunity to send a second copy of the invoice.	我們會再寄一份發票給您。
I have faxed again the copy of our invoice No.223 to you this morning.	今早我已再次傳真 223 號發票的副本給您。
We would like to know when you will settle the balance of $5,000.	我方想要知道您何時能付清五千元的餘款。
We would ask you to fulfill this obligation at your earliest convenience.	希望您儘早履行付款義務。
Please submit payment through your usual account.	請以您常用的帳戶匯款給我們。
You can remit to the Bank of Taiwan.	您可以匯款到台灣銀行。
You can pay by company check.	您可以用公司支票付款。

展現強硬的態度

According to the contract, you have to settle the account within 20 days after you received the shipment.	根據合約，您必須在收到貨物後的二十天內付款。
According to the terms of the seasonal account agreed upon, we would like to receive settlement by 15th April.	根據雙方同意的每季合約條款，我方希望能在四月十五日前收到款項。
According to our terms, you should pay by the end of May.	根據付款條件，您應該於五月底前付款。
As your payment is considerably overdue, we must ask you to remit the sum immediately.	您的款項已經逾期很久了，我方必須要求您立即付款。
Please do it without any delay.	請勿有任何延遲。

We ask you to settle the account immediately.	我方希望貴公司立即付款。	**Part 1** 入門篇
Please make remittance within the period agreed.	請於規定的期限內付款。	**Part 2** 求職篇
Please make an immediate payment to our bank account.	請儘速匯款至我方的銀行帳戶。	
Your settlement is already one month overdue.	您的貨款已逾期一個月了。	**Part 3** 客戶往來篇
The payment was due on May 31.	支付期限為五月三十一日。	
If you cannot settle the account immediately, we will be obliged to take legal steps.	如果您無法立即付款，我方將採取法律行動。	**Part 4** 人際互動篇
All orders will be shipped upon receiving the remittance.	確認匯款成功後，產品即會寄出。	
If you still have any queries unresolved regarding the invoice, please let us know.	如果您對帳單還有任何疑問，請讓我們知道。	
If there are other reasons we should know, please inform us.	若有其他原因，請通知我們。	
As always, please contact me if you have any concerns.	一如往常，如有任何問題，請隨時與我聯繫。	
Please confirm if you can meet the deadline, and if not, please contact us.	請確認您是否能準時付款，如果不行，請告知我方。	
We should be pleased to receive your check at your early convenience.	我方將會樂意收到貴公司提前寄給我們的支票。	
We hope to hear from you soon about the remittance.	我們希望很快就能收到您已匯款的通知信。	

If you can make remittance at your earliest convenience, we will be obliged.	如您能提早匯款，我們將不勝感激。
We look forward to receiving your remittance within five days.	我們希望在五天內收到您的匯款。
We believe you will send the remittance soon.	我們相信您會盡快匯款。
We would ask your prompt clearance of all invoiced amount.	我們希望您能儘速清償所有貨款。
We ask you to settle the account immediately.	希望您立即付款。
Regarding invoice No.2468, please make full payment as soon as possible.	關於 2468 號發票的款項，還請儘速付清。
Your urgent attention to this issue would be highly appreciated.	若您能優先處理此事項，我們將非常感激。

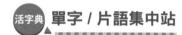

實力大補帖 Let's learn more!

活字典 單字 / 片語集中站

📍 **催款與延遲**

invoice 付款通知	**payment** 支付款項
due date 到期日	**overdue** 逾期的
overdue payment 逾期款項	**overdue amount** 逾期金額
overdue by...days 逾期…天	

Part
1
入門篇

Part
2
求職篇

Part
3
客戶往來篇

Part
4
人際互動篇

8-08 第二次催款信

在第二次的催款信中，語氣可以稍微強烈些，但因為並未掌握對方逾期的原因，所以可於信末加上「如果您已經付款，請忽略此信件」等類的敘述會比較恰當。

From:	Terry Chang
To:	Mr. Miller
Subject:	2nd Payment Reminder (Order No.991AL110)

Dear Mr. Miller:

Two weeks ago, we have reminded you of the outstanding balance of $5,000. So far, we still have not received your check.

Please make your payment as soon as possible, and advise the reason of delay in payment.

If you have sent the payment, please ignore this mail.

Best regards,

Terry Chang

米勒先生您好：

我們已於兩週前提醒您有關尚未付清的五千元餘額。截至目前為止，我們仍未收到您的支票。

請儘速付款，並告知延遲的原因。

如果您已經付款，請忽略此信件。

張泰瑞 謹上

信件開頭

In our mail three weeks ago, we directed your attention to the overdue payment of $2,000 in May.	我們三週前曾去函，提醒您五月份的兩千元款項尚未付清。
I am writing to you again to remind you that the balance $3,500 is still outstanding.	再次提醒您：帳單仍積欠三千五百元。
We have written to you on May 1st concerning our April statement, which is still outstanding.	我們已於五月一日寫信通知您：四月份的結算金額尚未繳清。
On April 8, we wrote to alert you that we had not received payment for your order.	我們已於四月八日寫信通知您：本公司尚未收到您的訂單款項。
This is the second mail regarding your account which has not been cleared.	針對您欠款的信件，這是第二封了。
In our last mail, we have reminded you about your unpaid account.	我們在上封郵件中提及您未繳的款項。
We are still waiting your payment for order number 5287.	我們仍在等候 5287 號訂單的結款。
20 days have passed; we still have not received your reply.	已經過了二十天，但我們仍未收到您的回覆。
In your reply to my mail, you promised you would clear the account yesterday.	您在回函中曾承諾，昨日就會結清帳款。
According to our bank, the account has not been settled yet.	根據我方銀行的說法，您仍未付款。

要求對方準時付款

- Please make your payment by 13th April.　　請於四月十三日前付款。

- Please send a check to clear the amount.　　請寄支票來結清帳款。

- We hope you will not withhold the settlement any longer.　　希望您不要再延遲付款。

緩和語氣的講法

- Your previous payment record shows your sincerity to keep your account up-to-date.　　您先前的付款紀錄顯示，您總是很有付清帳款的誠意。

- We therefore assume something has happened to make it difficult to make your payment.　　於是我們猜測可能發生了什麼導致您難以付款的事。

- We have not urged for the payment as you have a good record.　　由於您的良好紀錄，我們從未催討過款項。

- Please let us know what happened and perhaps we can provide a solution.　　請知會我們，也許我們能提供解決方法。

- We don't know what might be causing the delay in payment.　　我們不清楚導致延遲付款的原因。

- If you would like to discuss other payment options, we would be happy to oblige.　　如果您想要討論其他的付款方式，我們很樂意配合。

- If you have any difficulties, please let us know.　　如果您有任何困難，請告知我方。

- If there is some special reason for the delay in payment, we would welcome an explanation.　　如果此次的延遲付款有任何特殊原因，我們很樂意了解。

8-09 第三次催款信

第三封催款信，態度須更為強烈，告知對方經多次催詢仍無回應，甚至可提及將採取法律行動。但還是要提醒對方仍有妥協機會，請對方務必展現出誠意。

From:	Terry Chang
To:	Mr. Miller
Subject:	Notice to Pay (Order No.991AL110)

Dear Mr. Miller:

We have reminded you of the outstanding **balance**[1] by many mails, but have not received any reply or **remittance**[2] from you yet.

If we do not receive your **response**[3] within the next three days, we will have to **hand**[4] this matter to our **lawyers**[5]. We would be very sorry to have to take such **legal**[6] action after a long connection with your company. We sincerely hope this will not become necessary.

Best regards,

Terry Chang

米勒先生您好：

關於您尚未結清款項一事，我們已多次寫信給您，但卻仍未收到任何回應。

若接下來的三天內，我方仍未接到回應，就會將此事交由律師處理。我們已合作這麼長的時間，若真的必須訴諸法律，我們將感到非常遺憾，真心希望毋須採用這種方式解決。

張泰瑞 謹上

key words ▲ E-mail 關鍵字一眼就通

1 balance 名 結餘　　**2** remittance 名 匯款　　**3** response 名 答覆

4 hand 動 給；傳遞　　**5** lawyer 名 律師　　**6** legal 形 法律上的

複製 / 貼上萬用句 Copy & Paste

信件開頭

We have written you two mails to ask you to clear the balance.	我們已經寫了兩封信請求您繳清餘款。
We have sent you our statement for your May account twice.	我們已寄給您兩次五月份的結算單。
We wrote to you on April 3 and again on April 17.	我們分別於四月三日以及十七日寫信給您。
You have not responded to our earlier two mails about the overdue settlement.	您仍未回覆我方先前兩封有關逾期帳款的信件。
We have not received replies to our two previous mails of May 1 and May 12 for payment.	我們於五月一日和十二日寄出兩封繳款郵件，但至今仍未接到任何回覆。
After two reminders, we still have not received either your remittance or a delay reason from you.	經兩次提醒後，我方仍未收到您的款項，您也沒有告知拖欠的理由。
We have made several attempts to contact you by telephone to remind you of the payment.	我們已多次以電話提醒您付款。
Neither a reply nor the remittance from you is received.	我方沒有收到您的回覆或是匯款。
Two months have passed, and your account balance is still outstanding.	兩個月已經過去，而您仍未付清剩下的帳款。
You still owed us the sum of 5,000 pounds of the invoice No.55564.	根據 55564 號發票，您仍然欠我方五千英鎊的帳款。
Your balance of $3,500 for order No.22569 is now more than 45 days overdue.	編號 22569 號的訂單仍有三千五百元的餘款未付清，且已逾期四十五天。

以貨品為切入點

We believe that the goods we shipped to you were satisfactory and our records are in order.

我們相信寄去的貨品令您滿意，而本公司的紀錄也無誤。

If there are any problems with the shipment or the invoice, please contact us as soon as possible.

若貨運或發票有任何問題，請儘早通知我方。

If there is not any problem with the goods, you must pay the amount in full immediately.

如果貨品沒有問題，請您立即付清款項。

緩和的語氣 & 解釋理由

We always esteem you as a reliable customer.

我們一直認為您是可信任的客戶。

We always do our best to support you in all respects.

我們總是盡全力支持您。

We believe that your credit and reputation are still good.

我們相信您的信用及聲譽依然良好。

You are not only an important customer, but also a friend for us.

您不只是一位重要的客戶，更是我方的朋友。

We hope you would at least explain why the account continues to be unpaid.

我們希望您至少能解釋至今仍未付款的原因。

At least please offer an explanation for the delayed payment.

至少請您告知延遲付款的原因。

We have plans for investment recently; therefore, we also have some difficulties on financial management.

我們近日有投資計畫，因此我方在財務的調度上也有些困難。

We are experiencing some financial difficulties.

我們目前正面臨財務困境。

To preserve your credit standing, can you please settle your payment within the next three days?

為確保您的信用，可否請您三天內付清款項？

強硬的態度

We are disappointed that we have not received any response from you.

我們很失望仍未接到您的任何回覆。

This is the third and final reminder for the payment.

這是第三封，也是最後一封催款通知。

We are unhappy that we still got no response from you.

由於您遲遲未回覆，讓我們很不高興。

We are unhappy that the amount of $5,000 is still outstanding.

我們不滿意五千元的貨款至今仍未付清。

You are given enough time to clear the balance.

您已獲得足夠的付款時間。

The payment is seriously delinquent for long.

這筆款項已經嚴重拖欠。

Please try your best to settle this matter urgently.

請您立即解決問題。

Please give us an explanation, and settle your account ASAP.

請告知您未付款的原因，並儘速結清帳款。

Please don't keep silent anymore.

請勿再沉默。

Please understand the seriousness of this matter.

請了解這件事的嚴重性。

Our records show that you have always delayed the payment this year.

我們的紀錄顯示，您今年總是延遲支付貨款。

Unless the explanation is satisfactory; otherwise, we cannot allow the amount to remain unpaid.

除非有合理解釋，否則我方無法允許賒欠帳款。

We insist on payment within the next three days.	我們堅持此筆款項必須於三天內繳清。
Please do not delay any longer and please contact us.	請勿再拖延，並與我方聯繫。
Since you have not paid, we will have to cancel your order.	由於您未付款，因此我方必須取消訂單。

法律途徑：提供轉圜餘地

We feel that we have shown our patience and treated you with consideration.	我方認為，我們已經很有耐心，也考量過您的立場。
We have to take this case to court.	我們會訴諸法院。
If we take the legal step, you may be liable for the legal expenses.	如果我們訴諸法律，您可能要負擔訴訟費用。
We hesitate to do this currently.	我們現在仍未決定是否要這麼做。
You should know if we take legal actions, it will damage your standing in the community.	您應該知道，如果我方採取法律行動，將會損害貴公司在本區的聲譽。
The legal actions might damage your credit standing.	法律訴訟可能會損害您的信用。
We hope you will send us the full settlement by May 20.	我們希望您能於五月二十日前付清款項。
After the date, we will proceed with the above-mentioned actions.	過了這個日期，我們就會採取上述行動。
We will hand this matter to our lawyer.	我們會交由律師處理此事。
We hope such unpleasant condition can be avoided.	我們並不想要造成雙方的不愉快。

We sincerely hope that we can avoid the unpleasantness of taking legal actions.

我們衷心希望能夠避免令人不快的法律途徑。

We are willing to overlook this delinquency if you can settle the account due by May 10.

如果您於五月十日前付款，我們就不會再追究這次逾期的責任。

法律途徑：下最後通牒

We will be compelled to take legal actions unless we receive your remittance in the next five days.

若五天內未收到匯款，我方將會採取法律行動。

If not paid by December 31, we will have no choice but to place the matter in other hands.

如果您沒有在十二月三十一日前付款，我們就只好採取其他行動。

We will be forced to hand this matter to our attorney.

我們就只好將此事交由律師處理了。

We have no choice but to turn the matter over to our lawyer.

在沒有其他選擇的情況下，我們也只好將此事交由律師處理。

Without any response from you, we will be compelled to turn the account over to our lawyers.

若您不回覆此信，我們就必須把此事移交給律師處理。

Please be informed that our legal department intends to file a suit against you for collecting the payment of invoice No.552.

此信通知您：我方法務部門將訴諸法律途徑，收取 552 號發票的款項。

Unless we receive your check for $4,500, your account will be placed in the hands of our attorneys for collection.

除非我們收到您四千五百元的支票，否則該筆款項將交由我方律師託收。

If you still keep silent, the next mail you receive will be from our lawyer.

要是您仍選擇保持沉默，那我方將會請律師寄下一封通知給您。

We must regrettably inform you that your invoice No.552 has been forwarded to our collection agency.	我們很遺憾在此通知您：552號發票的收款事宜已交給我方的收帳公司了。
The matter of your overdue account has been turned to a collection agency.	您逾期繳款一事，已交由我們的收帳公司處理。
The collection agency will begin the litigation process to recover the amount due.	收帳公司將採取法律行動追回帳款。
We intend to take legal actions to enforce payment.	我們打算訴諸法律強制您履行付款行為。
We regret to take legal actions to recover payment.	我們對於要訴諸法律強制償付的行動感到遺憾。

信件結語

We are prepared to give you a further opportunity.	我方仍希望再給您一次機會。
Please help us to avoid other legal actions.	請勿讓我方採取法律行動。
Please give the matter your immediate attention.	請務必關切此事。
Please instruct your bank to make remittance within three days.	請指示您的銀行於三天內付款。
Please telephone me and make an effort to show that you are willing to meet obligations.	請致電給我，並展現出貴方願意付款的誠意。
We are writing to you in a final effort to ask you to clear your account.	我們已盡力，這是最後一封要求您付款的信件。
We are now making a final request for payment.	這是我們最後一次向您提出付款要求。

8-10 匯款延遲致歉

撰寫道歉信件時，態度務必誠懇。另外，直接誠懇的道歉會比絮絮叨叨找一堆理由搪塞來得有效，切勿拐彎抹角。

From:	Betty Wang
To:	Rick Taylor
Subject:	Apologies for Delayed Payment (Order No.3991)

Dear Mr. Taylor:

Due to the economic downturn, our sale of books has dropped. We are very sorry for the 5-day late payment. However, we still hope you can understand our situation.

We ensure that you will receive the remittance on time in the future. Thanks again for your understanding.

Best regards,

Betty Wang

泰勒先生您好：

由於經濟蕭條，我們的書籍銷售量因而驟減。逾期了五天才付款，對此我們深感抱歉。然而，我們仍希望您能體諒我方的困難。

我們向您保證，日後都會準時付款。再次謝謝您的諒解。

王貝蒂 敬上

信件開頭：致歉

We have received your mail reminding our overdue payment.	我們有收到您提醒我們款項已逾期的通知。
I am sorry that we forgot to make payment.	很抱歉，我方忘記付款。
We are so sorry for remitting you so late.	我們很抱歉這麼晚才付款。
We are terribly sorry for the late remittance.	很抱歉這麼遲才匯款。
My sincere apologies for the delay.	誠摯地為此次延誤致歉。
As I was sick and absent from the office, it is my failure not leaving instructions to pay the account.	因為我生病請假，未交代他人支付這筆帳單，這是我的疏失。

回覆 & 解釋

Our accounting department is looking into the matter.	本公司的會計部門正在清查原因。
The check was sent out on April 4, and should have reached you by now.	支票已於四月四日寄出，應該已經寄達才對。
I am terribly sorry for the delay on payment caused by the recent devaluation of our currency.	很抱歉，由於近日貨幣貶值，導致我方延遲付款。
We intended to clear the account before now.	我們本打算付清。
Due to the economic depression, our customers have not settled their accounts as well.	因為景氣低迷，所以我們的客戶遲遲未付款。

🖱	We have been subject to a number of late payments from our customers.	本公司常受許多客戶延遲付款之苦。
🖱	Our goods have not been sold out.	我方貨品仍未售完。
🖱	Our computers have some serious problems.	我們的電腦出了一些嚴重的問題。
🖱	We have tried our best to overcome this difficulty.	我們已經盡力克服困難。
🖱	In order not to omit your invoice, please mail all invoices to the mail address of our accountant from now on.	為了不遺漏您的發票，往後的發票請寄至我們會計的電子信箱。

信件結語

🖱	There is no excuse for this.	對此沒有任何藉口。
🖱	We have remitted US$5,000 by T/T today.	我們今天已電匯五千美元。
🖱	Such delay will not happen again.	這樣的延遲不會再發生。
🖱	I promise that this will be the last time I ever put myself in such situation.	我承諾這種狀況將是最後一次發生。
🖱	I apologize again for the delay.	再度為此次延遲向您致歉。
🖱	Please accept our sincerest apology for the late remittance.	請接受我方對此次延遲付款誠摯的道歉。
🖱	Thank you very much for your understanding.	感謝您的諒解。
🖱	We are very satisfied with the goods you sent us in the last order.	我們非常滿意上一批貨品。
🖱	We hope to place another order with you in the very near future.	我們希望近日能夠再度與您合作。

撰寫此類信件時，應鄭重地向對方道歉，並簡述理由，若能在回信時就已經付清款項最好。

From:	Victor Watson
To:	Ms. Zheng
Subject:	Re: Payment Reminder (Order No.991)

Dear Ms. Zheng:

We have received your mail informing us that we have not **cleared**[1] the balance. Upon checking the cause of the delay, we found it had not been paid due to an **oversight**[2] on our part.

We have sent our check for $1,500 this morning. It will reach you within two days. We are very sorry to keep you waiting. Please accept our **sincerest**[3] **apology**[4] for any inconvenience this matter has caused you.

We hope you understand that we have no **intention**[5] to delay the payment. And this will not **happen**[6] again next time.

Best regards,

Victor Watson

鄭女士您好：

我們已接到您通知我方未付清帳款的郵件。經過調查後，我們發現此次延遲是由我方的疏失所致。

今天早上我們已經寄出一千五百元的支票，您會在兩天內收到。很抱歉讓您久等，造成您的不便，還請接受我方誠摯的道歉。

希望您能了解我們並非故意要延期付款，日後我們會避免類似的情形再度發生。

維克多‧華生 敬上

E-mail 關鍵字一眼就通

1 clear 動 償清

2 oversight 名 疏忽出錯

3 sincere 形 衷心的

4 apology 名 道歉

5 intention 名 意圖

6 happen 動 發生

複製 / 貼上萬用句 Copy & Paste

信件開頭：致歉

I am replying your mail of May 20 requiring us to settle our payment due on May 1st.	此函回覆您五月二十日要求我們付清五月一日款項的郵件。
I apologize for the delay in payment.	我很抱歉延遲付款。
I apologize for the late payment.	我為延遲付款致歉。
We are deeply sorry for not clearing the account sooner.	我方對拖延帳款一事深感抱歉。
Let me apologize first for not settling the account.	有關尚未付款一事，請先接受我的道歉。
We are sorry we have not yet cleared the statement for $5,000.	很抱歉我們尚未付清五千元的帳款。
We were surprised to receive your mail in which you mentioned that you had not received our check.	我們很訝異您來信提及未收到我方支票的事。
We checked our account books immediately after receiving your mail.	收到您的郵件後，我們馬上就檢查了帳本。
We checked our records immediately after receiving your mail.	收到您的郵件後，我們立即調查紀錄。

解釋延遲付款的原因

Please allow me to explain the reason for our overdue payment.	請讓我向您說明我們延遲付款的原因。
Our accountant was sick last week. That was the reason why we were unable to settle the account.	我們的會計師上週生病，所以才無法付款給您。
One of our customers went bankrupt; therefore, we are facing financial problems now.	我們的客戶破產，因此我方目前正遭遇財務困難。
We have experienced a serious situation of non-payments by our suppliers.	我方遭遇供應商均未付款的嚴重問題。
Would you please give us another week to settle the account?	可否請您再寬限我們一週的付款時間？
We would appreciate if you will give us a little more time to settle the account.	若能再多給我們一點時間清償款項，我方將非常感激。
We will try our best to send you either entire or partial payment before June 20.	我們會盡力在六月二十日前寄出部分或全部貨款。

先前已經付款

Our book shows that we have paid you $2,500 by bank remittance on May 20.	我方的帳目顯示：我們已於五月二十日透過銀行匯款兩千五百元給您。
We have instructed our bank to remit the balance some time ago.	我們之前已指示銀行匯寄餘款給你。
We will fax you a copy of appliance sheet of remittance.	我們會傳真銀行提交的匯本申請單影本給您。
We don't understand why you haven't received our remittance until now.	我方不清楚為何您仍未收到匯款。
Maybe the bank has not informed you yet.	銀行也許尚未通知您。

Please check with your bank.	請向您的銀行確認。	Part 1 入門篇
Our check was posted to you on November 3.	我們已於十一月三日將支票寄給您。	
You should be able to confirm the receipt of the money in your account on April 6.	您應該會在四月六日收到帳款。	Part 2 求職篇
We are afraid that the check might be lost.	支票恐怕已經遺失。	
We have instructed our bank not to pay on the check.	我方已指示銀行止付。	Part 3 客戶往來篇
A replacement check will be sent to you soon.	我們會儘速寄新的支票給您。	

快速採取行動

The remittance procedures took longer than we expected.	匯款程序所需時間比我們想像的還要長。
I have checked our records and found our oversight.	我已檢查我方紀錄，發現是我們漏看了。
We will correct the mistake.	我們會馬上更正錯誤。
We have asked our accountant to settle the account to you.	我們已請會計師付款給您。
We have asked our accountant for an immediate payment to you.	我們已要求會計立即付款給您。
We found that we made an oversight in sending the check to you.	我方在寄送支票上有所疏失。
The sum of US$5,000 will be sent to you by T/T.	五千元美金的貨款將會電匯給您。

🖱	Our overdue payment will be remitted to your account today.	我方已逾期的款項會於今日匯入您的帳戶。
🖱	A bank draft will be sent to you.	我們會寄銀行匯票給您。
🖱	We have posted a check for $5,000 by registered mail.	我們已經用掛號寄出五千元的支票。
🖱	Please be patient and wait until next Monday for the payment.	煩請耐心等候，下週一您就能收到貨款。
🖱	We are sure that the remittance will be arrived tomorrow.	我們保證您明天就能收到匯款。

信件結語

🖱	Your patience is highly appreciated.	非常感謝您的耐心等候。
🖱	Please let us know if there is any problem.	若有任何問題，請告知我方。
🖱	We look forward to working with you again in the near future.	我們希望不久後能夠再次與您合作。

🛩 實力大補帖 Let's learn more!

深入學 資訊深度追蹤

📍 表「請求」的單字

01 ask（請求准許）：最一般的用語，不管對象是誰都能使用，常與 for 連用。

02 request（請求）：很正式的用語，有禮貌地向對方懇求。

03 require（需要）：因某種條件而提出的要求。

04 claim（要求）：通常為「根據某種權利」而做的要求。

05 demand（要求）：語氣強烈地要求對方，使用時須特別注意。

Part **4**
人際互動篇

拉近彼此的關係

 本章焦點 Focus！

| ✔ 同事間的溝通 | ✔ 與客戶的往來 | ✔ 公司的通知 |

　　本章以「個人」為主體，教你用 E-mail 拉近與同事 & 客戶的關係。如何把公事交辦得巧，讓同事樂於與你合作；善用信件噓寒問暖，抓住客戶的心，也守住自己的業績。除此之外，Unit 6 還會教你如何以「公司」為主體，通知各種事項。

辦公室點滴
At the Office

1-01 辦公室成立通知

在「辦公室成立通知」的信件中，應提及辦公室開幕的時間，除此之外，還可以提供網址，讓對方直接點選連結，以查詢更多資訊。

From:	Lisa Chen
To:	teachers.all@mail.net
Subject:	Opening of LIC

Dear English Teachers:

I'm very glad to inform you that The Language **Immersion**[1] Corner (LIC) in our **Applied**[2] English Department is going to open at 12:20 p.m. on Monday, September 10.

Here I would like to let you know the main **purpose**[3] of LIC. It **aims to**[4] provide our students with a comfortable environment to **facilitate**[5] their language learning. In LIC, there are several **cozy**[6] corners for you to sit back and relax before or after your teaching hours. Students are also excited about the opening of LIC and looking forward to chatting with our English teachers there.

We would also appreciate your ideas and comments on how to make good use of LIC and make it better. We are looking forward to seeing you at LIC.

Best regards,

Lisa Chen

親愛的英文老師：

很開心通知您，應用外語學系的語言學習中心即將於九月十日（星期一）中午十二點二十分開幕。

語言學習中心的主要目的為提供舒適的語言學習環境。這裡有許多舒適的角落，可讓您於上課前及下課後小憩。學生們對於語言學習中心的開幕感到非常興奮，也很期待在這裡與英文老師對談。

若您能惠賜對語言學習中心的意見，以及改善本中心之建議，我們也會非常感激。非常期待能在這裡與您相見。

陳麗莎　謹上

key words ▲ E-mail 關鍵字一眼就通

1 immersion 名 沉浸　　　2 applied 形 應用的　　　3 purpose 名 目的

4 aim to 片 旨在　　　5 facilitate 動 促進　　　6 cozy 形 舒適的

複製 / 貼上萬用句　Copy & Paste

We are pleased to inform you that Courage Company located on Yarmouth Avenue is now open.	很開心通知您，勇氣公司已於雅茅斯大道開幕。
Effective as of September 28, LIC will provide students a comfortable environment to learn English.	從九月二十八日起，語言學習中心將會提供學生學習英文的舒適環境。
Foreign Language Center (FLC), founded in August of 2018, is a newly established unit.	外語中心創立於二〇一八年八月，是個新成立的單位。
We hope that you will come and give us your advice and opinions!	歡迎您的蒞臨與指教！
For more information, please refer to our homepage: https://www.lic.edu.tw.	欲了解詳情，請上我們的網站：https://www.lic.edu..tw。

1-02 會議通知

公司內經常會需要寫「會議通知」的信函，這類信件的重點在清楚，所以記得於信中提及正確的開會時間和地點，信末可再提醒收件者「務必準時出席」。

From:	Patrick Jones
To:	all@mail.net
Subject:	Sales Meeting on May 5

Dear Managers:

Please be informed that a manager's meeting for the sales strategy will be held at 9:00 a.m. on Friday, May 5, in our Meeting Room 2.

Please make sure your presence.

Best regards,

Patrick Jones

諸位經理：

謹通知，公司將於五月五日早上九點鐘，於第二會議室針對銷售策略舉行主管會議。

請務必出席會議。

派翠克‧瓊斯 謹上

複製 / 貼上萬用句 Copy & Paste

宣布開會時間 & 地點

 **There will be a meeting about sales promotion at the head office next month.** 總公司預計於下個月召開行銷會議。

Part
1
入門篇

Part
2
求職篇

Part
3
客戶往來篇

Part
4
人際互動篇

The meeting will be held in the Conference Room No.5 at the head office.

會議將於總公司的第五會議室召開。

The next Monthly Management Meeting will be held at 10:00 a.m. on Friday.

每月的主管會議將於週五早上十點舉行。

The monthly meeting has been postponed to May 31.

本月例行會議延期至五月三十一日舉行。

A manager's meeting will be held at 4 p.m. next Monday, August 23, in the meeting room.

經理級會議將在下週一（七月二十七日）下午四點鐘，於會議室舉行。

Our training on the new computer software will be given on Friday, May 5.

新電腦軟體的訓練課程定於五月五日，星期五。

The meeting will be held in our Meeting Room No.1 at the office.

會議將於公司的第一會議室舉行。

We'll hold the meeting from 2:30 p.m. to 5:00 p.m. on Tuesday, Nov. 12 in meeting room 202.

我們將於週二下午兩點半至五點舉行會議，地點在 202 號會議室。

There is a meeting to discuss marketing issues at 6 p.m. Tuesday, May 29 at ABC Tower.

五月二十九日（週二）下午六點，在 ABC 大樓將有一場討論行銷議題的會議。

Please mark your calendar accordingly.

請依此安排您的行程。

點出開會目的

The purpose of the meeting is to discuss the design of our new products.

開會目的在於討論新產品的設計。

The purpose of the conference is to discuss the international marketing strategy in Europe.

會議目的是要討論歐洲國際市場的行銷策略。

Our primary focus of the meeting will be product promotion strategies in the China market.	開會最主要的目的在討論中國市場的行銷策略。
In the conference, we will discuss issues about our new office in Shanghai, China.	席間我們將針對成立於中國上海的新辦公室做討論。
We need to have a meeting to discuss the withdrawal from the Europe market.	我們有必要針對退出歐洲市場一事開會討論。
We would like to invite you to a meeting about how to use the Apple software.	我們想要邀請您來參加一場會議，學習如何使用蘋果電腦的軟體。
There will be a top sales' meeting about sales promotion skills at the head office next week.	下星期在總公司將會舉行頂尖業務的會議，教您成為推銷達人。

與會議相關的提醒

Please do some homework by reading the following reports in advance.	請事先閱讀以下的報告。
As the details are attached, please read through the files in advance.	附上相關訊息，請詳讀。
Be sure that you are ready with questions and comments.	請您準備好問題與建議。
Please come and prepare with your questions and suggestions.	出席時請準備好相關問題與建議。
The following is the tentative agenda.	以下為初步開會議程。
Following is the agenda for the meeting:	以下為議程：
If there are other topics you'd like to add, please let me know.	若想增加其他議題，請讓我知道。

If there are any issues you would like to place on the agenda, please let me know via e-mail by Wednesday.

如有其他想討論的議題，請於週三前寄電子郵件給我。

要求對方出席

We hope you will participate in the conference.

我們希望您能出席會議。

Please make sure that you will be present.

請確認您能出席。

Your presence is requested.

請您務必出席。

You are required to be present.

請您務必出席。

It is imperative that you attend the meeting.

您必須出席這場會議。

All of you are required to be present at the meeting.

各位都必須出席本次會議。

Your attendance at the meeting this Thursday is important to us.

我們慎重地邀請您參加本週四的會議。

Please send someone on your behalf if you are unable to attend the conference.

如無法出席，請派代理人與會。

If you are unable to attend, please let me know who will be attending from your department.

如果您無法參加，請告知貴部門將由誰代表出席。

Please contact us if you cannot attend the meeting.

若您無法參加會議，請與我們聯繫。

If you are unable to come, please let me know by next Thursday.

若您不克前來，請於下週四前讓我知道。

1-03 請對方更改開會時間

Dear Judy,

Thank you for your mail regarding the monthly meeting in June. Unfortunately, I regret to inform you that due to some pre-arranged schedule, I'm afraid the meeting should be rescheduled.

May I suggest postponing the meeting to June 29 at 3:00 p.m.? I will give you a call to discuss and confirm the time. Sorry for the inconveniences caused.

Thank you for your patience and understanding.

Best regards,

Kevin

茱蒂您好：

謝謝您寄的六月份月會通知。不巧的是，因為我先前已排好行程，恐怕會議得改期。

是否能將會議延期至六月二十九日的下午三點呢？我會再打電話與您討論，並確認時間。抱歉造成困擾了。

謝謝您的耐心與體諒。

凱文 謹上

複製 / 貼上萬用句 Copy & Paste

I'm afraid I have to request to reschedule our meeting of June 15.	我恐怕必須更改六月十五日的會議時間。
Unfortunately, I will have to travel to Boston that week.	很不巧，我那週必須去波士頓。
May I suggest September 16 at 2:00 p.m.?	可以將會議更改至九月十六日下午兩點嗎？
I'm sorry for any inconveniences this has caused you.	造成您的不便，敬請見諒。
I'm waiting for your confirmation.	我等待您的確認。

實力大補帖 Let's learn more!

活字典 單字 / 片語集中站

📍 會議流程

agenda 議程	coffee break 休息時間
give a speech 演講	minutes 會議記錄
opening remark 開場白	preside 主持
reschedule 重新安排	session 會期

📍 會議中的物品

badge 會議中掛的名牌	microphone 麥克風
projector 投影機	pointer 簡報筆
reception 報到櫃台	screen 螢幕
whiteboard 白板	whiteboard pen 白板筆

1-04 活動通知

發送活動通知時，除了活動訊息之外，還可以於信末註明「請不參加的同事提前告知」，以便準備食材份量與活動內容。

From:	Winnie Liu
To:	all@mail.net
Subject:	Annual Dinner Party

Dear All:

We are having an annual dinner party at the National Hotel, Kaohsiung on Saturday night, January 20, from 6 p.m. to 9 p.m. Food and drinks will be provided. Formal dress is not required.

You are welcome to bring your family, but please let me know the number of people you're brining in advance.

If you cannot attend the party, please let me know as well.

Best regards,

Winnie Liu

親愛的各位：

我們將於一月二十日（週六）晚上六點到九點，於高雄的全國飯店舉辦年終晚宴。會提供飲料和食物，可著便服。

歡迎攜家人前來同樂，但請事先告知人數。

如不能參加，也煩請事先告知。

劉溫妮 謹上

通知活動

We are going to throw a year-end party for you.	我們將為您舉辦一場尾牙派對。
We will be chatting, dining and having fun.	大家可以一起聊天、用餐和玩遊戲。
Please come and enjoy the fellowship.	歡迎大家共襄盛舉。
We invite all our faculty, part-time teachers, and foreign teachers.	我們邀請所有職員、兼任教師與外籍老師一同來參加。

特殊事項

Please bring one gift and wrapped it nicely for gift exchange.	請準備一份包裝精美的禮物來交換。
You don't have to buy a new item; if you do, please control the price under NT $300.	不強制各位購買新品,若您覺得需要,價格請控制在新台幣三百元以內。
What you can do is to go through your shelf or closet for some present you received but has never been used, and bring it to the party.	您可以在書架或衣櫥裡找出別人曾經贈送,但您根本不曾用過的物品,並帶來交換。

信件結語

Feel free to bring your family or friends to the party.	歡迎攜伴參加。
If you plan not to attend the party, please inform us in advance.	若您不參加,煩請事先告知我們。
Please come and enjoy this year-end party together.	歡迎您前來參加尾牙,與我們同樂。

🖱 **Have a great weekend.**　　　　　　　祝您有個愉快的週末。

 實力大補帖 **Let's learn more!**

📍 服裝的質料

cashmere 喀什米爾羊毛	cotton 棉
denim 丹寧布料	leather 皮
nylon 尼龍	polyester 聚脂纖維
silk 絲	wool 羊毛

深入學 **資訊深度追蹤**

📍 美國常見的各種服裝要求（Dress Code）

01 White Tie / Ultra-formal：西方社交場合最正式的著裝（國宴或超正式的晚宴才會出現）。男性要穿燕尾服；女性則必須穿著下擺及地的長禮服。

02 Black Tie / Formal：最常見的正式服裝。男性必須穿西裝；女性的裙子則要過膝。

03 Semi-formal：半正式的服裝。雖然可以穿得輕鬆一點，但男性還是必須穿西裝；女性穿小禮服即可（但女性的禮服最好以緞面的為主）。

04 Business Causal：工作面試等場合適用的服裝，男性建議穿西裝；女性則為套裝。

05 Causal / Informal：可以隨便穿，但短褲、涼鞋還是必須避免。

From: Jack Lee

To: sales.all@mail.net

Subject: Notice of Absence

Dear colleagues:

I will be staying in Korea from next Monday to Friday for five days and be back to office on August 8.

Sorry for the inconvenience, and please reschedule your events accordingly.

Best regards,

Jack Lee

Sales Manager

親愛的同仁：

我下週會去韓國，週一到週五，共五天，八月八日才會回來上班。

很抱歉造成困擾，並煩請重新規劃您的行程。

銷售經理 李傑克 上

複製 / 貼上萬用句 Copy & Paste

通知休假 / 公休

We are informing you that ABC Company will be closed for the Chinese New Year holidays from Jan. 29 to Feb. 5, and will reopen on Monday, Feb. 6.

謹此通知，ABC 公司在農曆新年期間休假（一月二十九日至二月五日），二月六日重新開始上班。

It is our pleasure to announce that from this year, we will have a Labor Day off on May 1st.

很榮幸宣布，今年起五一勞動節放假。

Our store will be closed during the Easter holidays and will reopen on Wednesday, May 20.

我們的店在復活節期間將暫停營業，並於五月二十日（星期三）重新營業。

The store will be closed on Thursday night due to the annual dinner party.

星期四晚上為年終晚會，因此暫停營業。

The shop will be closed on Wednesdays and Sundays.

本店週三及週日公休。

The store will be closed on every Thursday night and Saturday all day from now on.

即日起，本店將於每週四晚間及週六全天公休。

For your information, our company just confirmed that Feb. 28 and Apr. 4 (both are Mondays) are official days off.

在此通知您，二月二十八日與四月四日（皆為週一），已確認為本公司的公休日。

信件結語

Please keep in mind it is "Holiday" these two days!

請記得：這兩天「不用上班」！

Please enjoy your Chinese New Year holidays.

請享受您的農曆年假。

Wish you all a wonderful holiday.

祝福各位假期愉快。

🖱 **Have a great week and Happy Holidays by then!** 　祝您有個美好的一週，並祝佳節愉快！

 **實力大補帖 Let's learn more!**

活字典 ▲ **單字 / 片語集中站**

📍 **台灣國定假日**

Chinese New Year 新年	**Spring Festival** 農曆春節
Tomb-sweeping Day 清明節	**Labor Day** 勞動節
Dragon Boat Festival 端午節	**Mid-autumn Festival** 中秋節

📍 **美國國定假日**

Good Friday 耶穌受難日	**Easter Holiday** 復活節假期
Memorial Day 陣亡將士紀念日	**Independence Day** 國慶日
Thanksgiving Day 感恩節	**Christmas Day** 聖誕節

深入學 ▲ **資訊深度追蹤**

📍 **與「休假」相關的英文單字**

01 holiday：通常用來指較短期或是為期幾天到幾週的休假。例如 Christmas 和 Chinese New Year。

02 vacation：長時間的休息。例如包含好幾個月的 summer vacation（暑假）。

03 day off / leave：兩者皆有「休假日」和「請假」的意思。例如：It was my day off. 我休了假。/ I took leave today. 我今天請假。

請假前，須寫信告知業務相關夥伴自己的休假起訖日，以及休假期間的職務代理人，以方便彼此作業。

From:	Lauren Hill
To:	all@mail.net
Subject:	Notice of Absence

Dear colleagues:

Please be informed that I will be out of office from Wednesday through Friday, May 5-7 due to the **business trip**[1] to Japan. In my **absence**[2], Ms. Jennifer Chang will be managing Marketing Department. She can be reached at jennychang@mail.net, or telephone: 0922-112-333.

During my trip, I will be connecting to the Internet in the hotel in order to **catch up with**[3] your program **progress**[4].

Please feel free to contact me by e-mail if you need my **assistance**[5] **urgently**[6].

Thank you very much!

Lauren Hill

Marketing Manager

親愛的同仁：

謹此通知，五月五日到七日（週三至週五），我會到日本出差。出差期間，將由珍妮佛小姐代理我的職務，處理行銷部門的業務。您可以寄電郵至 jennychang@mail.net，或打電話至 0922-112-333 聯絡她。

出差期間，我也會使用飯店的網路，以便確認各位的工作進度。

若急需幫忙，請用電子郵件連絡我。非常謝謝您！

行銷經理 羅倫‧希爾 上

Part
1
入門篇

Part
2
求職篇

Part
3
客戶往來篇

Part
4
人際互動篇

key words ▲ E-mail 關鍵字一眼就通

1 business trip 片 公差　　**2** absence 名 不在　　**3** catch up with 片 趕上

4 progress 名 進展　　**5** assistance 名 幫助　　**6** urgently 副 緊急地

🖅 複製 / 貼上萬用句 Copy & Paste

通知出差或休假

This is to inform that I will be out of office from May 1st to 5th.	謹此通知，我將於五月一日至五日出差。
I am now out of office on a business trip to Hong Kong from Oct. 10 to 16.	我目前正在香港出差，十月十日至十六日都會待在此地。
I am currently out of office with limited access to e-mail.	我目前出差，暫時無法上線收信。

告知職務代理人資訊

For urgent issues, please contact Mary at 07-2172-3563.	若有緊急事故，請聯絡瑪莉。電話為：07-2172-3563。
You can telephone him on 9242-3173 ext. 123 during business hours.	您可於上班時間打電話到：9242-3173，分機 123 與他聯繫。
Please e-mail him on paulwang@mail.net.	請寄電子郵件給他至：paulwang@mail.net。
E-mail me at brown.chen@mail.com.	請電郵給我至：brown.chen@mail.com。
If you want to discuss about the project, my skype account is tim@mail.com.	如果您想要討論企劃，您可以加我的 Skype 帳號：tim@mail.com。
Please e-mail me first before adding my skype account.	在加我的 Skype 帳號之前，麻煩先寫電子郵件告知我。

1-07 歡迎新進員工

在「歡迎新進員工」的信件中，記得要提供收件者有關新進員工的相關資訊，如姓名及職稱等，以幫助雙方日後的業務聯繫。

From:	Carrie Wang
To:	Danny Lee; Sandy Yang
Subject:	Welcome Our New Colleague

Dear Danny and Sandy:

We have added a new face to our HR department this week. Please note Mr. Brown has been promoted to HR manager. He will work with you and your coworkers in your department.

We hope the upcoming meeting will give you a chance to meet our new personnel.

Sincerely yours,

Carrie Wang

丹尼和珊蒂您好：

本週人資部門增加了一位新面孔。請注意：布朗先生被晉升為人資主管，他將與您部門的同仁共事。

希望能在最近的會議上介紹新進人員給您認識。

王凱莉 謹上

複製 / 貼上萬用句 Copy & Paste

🖱 **I am pleased to announce an appointment.** 很高興向您宣布一項任命。

We are glad to announce the new employment of Bella Thomas.	我們很高興宣布聘請貝拉‧湯馬斯。	Part 1 入門篇
I am glad to announce that Tess will be appointed as our sales manager.	我很高興宣布：泰絲被任命為業務經理。	Part 2 求職篇
We decided to bring some new blood to the team.	我們決定為團隊注入新血。	
We are very happy to announce that Lauren Brown will be our sales manager effective from today.	我們非常高興地宣布：即日起，本公司業務經理將由羅倫‧布朗先生擔任。	Part 3 客戶往來篇
Tina has many years of experiences in the marketing area.	蒂娜有許多年的行銷經驗。	
We believe her experiences will bring invaluable asset to our company.	我們相信她的經驗將替公司帶來龐大的資產利潤。	Part 4 人際互動篇
Please welcome Mr. Taylor to our company!	請大家一起歡迎泰勒先生！	

實力大補帖 Let's learn more!

活字典 單字 / 片語集中站

📍 向新進人員介紹環境（Employee Orientation）

corporate culture 企業文化	development 發展
discipline 紀律	job description 工作說明
job specification 工作規範	objective 目標
orientation 情況介紹	performance 績效
policy 政策	training 培訓；訓練

1-08 職務升遷通知

在「職務升遷通知」中，首先可恭賀收件者獲得升職，再說明升遷的原因，最後可以給予鼓舞，希望對方未來能持續努力。

From:	Nancy Moore
To:	Eric Robinson
Subject:	Notice of the Promotion

Dear Eric:

It is our pleasure to inform you that you have been promoted to the challenging and **demanding**[1] **position**[2] of Factory Chief.

With all your hard work and **dedication**[3] towards your work, you have achieved this promotion within one year. We expect the same **endeavor**[4] from you in the future after this promotion. Your new salary structure and details about **compensation**[5] will be mentioned in the official promotional letter, which will be given to you very soon.

If you find any **queries**[6] or difficulties related to this matter, you can contact the human resources department.

Once again, many congratulations to you and all the best for your future growth.

Sincerely yours,

Nancy Moore

艾瑞克您好：

很高興通知您被晉升為工廠廠長，這將會更富有挑戰性。

因為您這一年的努力與付出，才有今天的升遷。我們期盼您升遷之後依然能盡心盡力地投入在工作中。近日會寄給您正式的升遷信函，裡面會列出您升遷後的薪資結構以及相關的津貼細項。

若有任何問題以或難處，可與人力資源部門聯繫。

再次恭喜您，也祝您的未來一片光明。

南西・摩爾 敬上

 E-mail 關鍵字一眼就通

1 demanding 形 高要求的　　2 position 名 職務　　3 dedication 名 奉獻

4 endeavor 名 努力　　5 compensation 名 津貼　　6 query 名 疑問

複製 / 貼上萬用句 Copy & Paste

You have therefore been promoted to Product Manager, effective from September 1st, 2018.	你被升為產品經理，從二〇一八年九月一日開始任職。
This promotion is in recognition of the great work you have done for this company.	因為您過去在公司的表現良好，因此獲得拔擢。
We are extremely pleased to inform you that the President has considered you suitable for the post.	我們非常開心地通知你，總裁認為你適合這個職位。
We are sure that you will meet the new responsibilities as a Factory Chief.	我們相信您會負起廠長的責任。
You are requested to hand over responsibilities of Sales Representative to Joan by August 31.	八月三十一日前，請務必將業務專員的職務移交給瓊安。
Congratulations on your achievement and we wish you all the best in the coming years.	恭賀您的成就，我們期待您未來的表現。

1-09 辭職通知

要在「辭職通知」中註明正式離職的日期，並感謝收件者於在職期間的照顧與配合。

From: Zoe Thompson

To: Ms. Morgan

Subject: Resignation

Dear Ms. Morgan:

I would like to inform you that I am **resigning**[1] from my position as Marketing Manager for Far East Company, **effective**[2] as of May 1.

Thank you for the **support**[3] and the opportunities that you have provided me during the last three years. I have enjoyed my **tenure**[4] here.

If I can be of any assistance during this **transition**[5], please let me know. I would be glad to help.

Kind regards,

Zoe Thompson

摩根小姐您好：

謹此通知您，我將於五月一日卸下在遠東公司的行銷經理職務。

謝謝您這三年來給予我的支持與機會，我在公司任職的這段時間很開心。

在這段過渡期間，若有任何需要幫忙的地方，請讓我知道，我很樂意提供協助。

柔伊‧湯普森 敬上

key words ▲ E-mail 關鍵字一眼就通

1 **resign** 動 辭職　　2 **effective** 形 生效的　　3 **support** 名 支持

4 **tenure** 名 任期　　5 **transition** 名 過渡期

告知離職的消息

Please accept this message as notification that I am leaving my position with Philips Company, effective as of September 20.

謹此通知您：我將於九月二十日離開飛利浦公司。

I am writing to notify you that I am resigning from my position as HR Manager with XYZ Company.

此函通知您：我即將卸下人力資源部門經理一職，離開 XYZ 公司。

I am writing to inform you of my decision to resign from ABC Inc., effective as of December 1.

在此通知您：我決定於十二月一日離開 ABC 公司。

The purpose of this letter is to announce my resignation from UMC Company, effective two weeks from this date.

此函目的為通知您：我即將在兩週後離開 UMC 公司。

I'd like to let you know that I am leaving my position at the company.

在此告知您：我即將離職。

My last day of employment will be May 30.

我最後一天的上班日為五月三十日。

對公司的感謝

I have learned a lot, and grown professionally during my time under your employ.

在您的帶領下，我學到很多，專業能力也成長不少。

Thank you for your understanding of my decision to leave the company, and all your support over the years.

謝謝您願意諒解我離開公司的決定，也感謝您這些年的支持。

I appreciate the opportunities I have been given at CTC, as well as your professional guidance and support.

很感謝 CTC 公司給我的機會，以及您給予的專業培訓和支持。

I have enjoyed my tenure at ABC and I appreciate having had the opportunity to work with you.	在 ABC 工作的這段時間我很開心，也感謝有與您共事的機會。
I have greatly enjoyed working with you for the past three years.	過去三年很高興能與您共事。
Thank you for the support and encouragement you have provided me during my time at the company.	謝謝您在我任職期間給予的支持與鼓勵。
I've enjoyed working with you and managing a very successful team.	很開心與您共事，並且管理一個非常成功的團隊。
I have enjoyed my tenure with the company.	我很享受在公司任職的時間。
The past eight years have been very rewarding.	過去八年很有價值。
I learned so much from my experience working at your firm, but I also felt it was time to move on to another position.	我在公司學到許多事情，但仍覺得是該邁向新階段的時候了。
The professional skills I have developed while working at your company will be extremely useful for my future career.	在公司所學到的專業技術對我未來的職涯非常有用。

提及自己的計劃

My illness will require extended treatment and recovery.	我的疾病需要長期的治療與休養。
I have accepted a position as Magazine Editor at a publisher in Taipei.	我已經接受台北一家出版社所提供的雜誌編輯一職。
I plan to return to school to pursue my interest in Computer Science.	我計畫重返學校，攻讀我喜歡的電腦科學。
On my part, this was not an easy decision to make.	對我而言，要下這個決定並不容易。

I feel it is time to move on to new opportunities and challenges.	我覺得該是把握新機會與挑戰的時候了。
This opportunity gives me the chance to grow professionally.	這個機會給了我在專業上進一步成長的契機。
Even so, I will miss all my colleagues, clients, and the company.	即便如此,我還是會想念這裡的同事、客戶以及這家公司。
I consider everyone I met here to be my friends, and I will miss you all.	這裡的每個人都是我的朋友,我會想念大家。

留下聯繫方式

Please keep in touch.	請保持聯絡。
I can be reached at mary.chen@mail.com or via cell phone 0927-282-331.	可以寫電子郵件到 mary.chen@mail.com 給我,或是致電 0927-282-331。
Please feel free to contact me with any questions about the projects I worked on.	若我處理的企劃有問題,請隨時與我聯繫。

給予祝福

Again, thank you so much for an excellent five years in Philips company with me.	再次感謝過去五年在菲利浦公司的共事時光。
I wish you all the best for your continued success.	希望您一切順利並成功。
I wish you and the company success in the future.	希望您與公司未來皆成功。
I wish you and the company all the best.	祝福您以及公司。

1-10 退休通知

「退休通知」為「即將退休者寄給同事的信函」。可於信中回顧往日的職涯,感謝同事,並可於信末表示希望與同事保持聯繫。

From:	Joseph Yang
To:	all@mail.net
Subject:	Notice of Retirement

Dear all:

The time has come for me to leave Philips Inc. I will be leaving the company at the end of June after 30 years of happy time.

I would like to sincerely thank you all for the **enjoyable**[1] working **relationship**[2] and wish you well for the future.

Thank you kindly for the retirement party and gifts. I shall certainly think of you all every time I play **golf**[3]!

Keep in touch.

Kind regards,

Joseph Yang

親愛的各位:

是時候要離開飛利浦公司了——一個我待了三十個快樂年頭的公司,而我即將於六月底離開。

衷心感謝能與你們共事,祝福你們未來一切順利。

謝謝你們替我舉辦了退休派對,還送禮物給我。以後我打高爾夫的時候,肯定會想起你們的!

記得保持聯絡。

喬瑟夫‧楊 謹上

Part
1
入門篇

Part
2
求職篇

Part
3
客戶往來篇

Part
4
人際互動篇

From:	Annie Lee
To:	all@mail.net
Subject:	Notice of Retirement

Dear All:

I would like to inform you that I will be retiring from my position with ABC Inc. as of the end of June.

I would like to take the opportunity to thank you for the professional working relationship we have enjoyed over the last six years.

I am looking forward to my retirement but will **surely**[4] miss working with you all. If I can be of any **assistance**[5] during the next few weeks, please don't **hesitate**[6] to contact me.

Sincerely yours,

Annie Lee

親愛的各位：

謹此通知：我將於六月底退休，離開 ABC 公司。

藉此機會感謝您們，過去六年來以專業的姿態與我共事。

雖然很期待退休生活，但肯定也會非常想念大家。接下來的幾週內，若有需要我協助的地方，請不用客氣，儘管聯絡我。

李安妮 敬上

 E-mail 關鍵字一眼就通

1 enjoyable 形 快樂的　　2 relationship 名 關係　　3 golf 名 高爾夫球

4 surely 副 無疑；一定　　5 assistance 名 援助　　6 hesitate 動 猶豫

通知：由退休者告知 & 感謝語

I would like to inform you that I am about to retire from my position as a sales manager from the company, effective as of Jan. 25, 2018.	在此通知您：我將於二〇一八年一月二十五日正式離職，自業務經理的職務退休。
I am writing to you this letter because I want to inform you that I am about to retire and I hereby give you a month's notice as required.	因應要求，謹於一個月前以此函通知我即將退休的消息。
I wanted to thank you for giving me the opportunity and privilege to work in your company.	感謝您賦予我在貴公司工作的機會。
I appreciate the things that the company has done for me, and I wanted you to know that I will treasure them for the rest of my days.	感謝公司為我做的一切，往後的日子裡，我會好好珍惜這些回憶。
I wanted to say thank you for the different opportunities you have given me.	感謝您給我許多不同的機會。
Your company helped me so much in developing my professional and personal growth.	公司培養我的專業，也讓我個人成長許多。
Truly, I have enjoyed working in your company, and I appreciate your sincerity to help me during my tenure with the company.	我真的很享受在貴公司的工作，也很感謝您對我的幫助。
I will miss everything from your company, especially with the bonds that I have established with the coworkers here.	我會想念公司的一切，尤其是與同事建立的情誼。

退休的理由

🖱 I have decided to take my retirement because I am not growing any younger.

由於我的年事已高，因此做出退休的決定。

🖱 I want to enjoy the very last moment of my life with my family and grandkids.

我希望能夠在晚年享受天倫之樂。

同事舉辦送別會之後

🖱 I am writing to thank you and all the staff at ABC Inc. for the party you held in my honor last week.

這封來信是要感謝 ABC 公司的各位同仁上週特地為我舉辦退休派對。

🖱 I also want to thank for the beautiful watch you presented me as a leaving gift.

也謝謝各位為我挑選了漂亮的手錶作為餞別禮。

🖱 As I said in my speech, I've really enjoyed my thirty-five years with the company.

如同我所說的那樣，這三十五年在公司，我真的過得很開心。

🖱 It has been a fantastic place to work and I enjoyed being part of such a dedicated team.

這裡有很棒的工作環境，我很開心能和這麼努力的團隊一起工作。

🖱 I certainly won't miss getting up so early in the morning, but I will miss being part of the team.

當然，我不會懷念必須早起的上班日，但我肯定會懷念與團隊一起工作的日子。

🖱 I look forward to seeing you all again sometime in the near future, perhaps for a round of golf or a wee drink.

期盼未來還能與大家相聚，無論是打場高爾夫還是喝杯小酒。

🖱 Thanks again for the most enjoyable evening.

再次感謝大家帶給我那麼愉悅的夜晚。

信件結語：歡迎聯繫

If I can be of any assistance, never hesitate to call me over the phone.

如果需要幫忙，歡迎隨時與我電話聯繫。

If I can be of any assistance, feel free to contact me with my number or e-mail.

如果需要幫忙，請隨時打電話或是以電郵聯絡我。

通知：他人的退休 & 感謝語

It is with regret that we have to inform you: considering his health condition, Mr. Tsai, our Vice President, will be retiring from active business on Jan. 1.

本公司副董事長蔡先生，由於健康因素，將於一月一日退休，特此告知，至感遺憾。

It is with sadness and emptiness that I am announcing the retirement of one of the foundations of our company, Mr. Thomas.

在此感傷地宣布，我們公司的其中一位創辦人，湯瑪斯先生即將退休。

His place as Director will be taken by Ms. Taylor.

他的處長一職將由泰勒女士接替。

Mr. Lee has been with the company for more than thirty years.

李先生在本公司服務已超過三十年。

During his tenure, the company experienced phenomenal growth.

他在職期間，本公司的業績有了驚人的成長。

People who have been under his supervision all reported his good nature and great management skills.

在他管理之下的同事們皆讚賞他敦厚的脾氣以及出色的管理能力。

Without him, there would have been many problems among some colleagues.

如果沒有他的話，同事間或許會產生許多問題。

Let us all wish him the best of luck on his retirement.

讓我們祝福他的退休生活順利。

Part
1
入門篇

Part
2
求職篇

Part
3
客戶往來篇

Part
4
人際互動篇

1-11 回覆退休通知

祝福他人退休時，記得對於退休者在職期間的工作表現給予讚揚，並祝福對方未來一切順利。

From:	Henry Wang
To:	Joseph Yang
Subject:	Re: Notice of Retirement

Dear Joseph:

Congratulations on your **retirement**[1]!

Thirty years have passed and now it's time for you to enjoy the next chapter in your life. We would like to thank for your **continuous**[2] hard work over the years, your **loyalty**[3] to the company and the **professional**[4] working relationship we had.

You will be **sorely**[5] missed and we wish you a long, happy and healthy retirement.

Sincerely yours,

Henry Wang

親愛的喬瑟夫：

恭喜您退休！

在職三十年，也該是時候向您人生的下一章節邁進了。我們感謝您不間斷的努力、對公司的忠誠、以及在工作上所展現的專業。

我們會想念您，也祝福您的退休生活健康快樂。

王亨利 敬上

1 retirement 名 退休　**2** continuous 形 連續的　**3** loyalty 名 忠誠

4 professional 形 職業性的　**5** sorely 副 很；非常

複製 / 貼上萬用句　Copy & Paste

得知對方退休的消息

Hereby I am confirming your notification of your retirement and relieving you from ABC Inc., effective as of Nov. 4, 2018.	在此通知：已核准您在 ABC 公司的退休，生效日為二〇一八年十一月四日。
Thomas, you have been with us for 30 years.	湯馬斯，您已經與我們一同工作三十年了。
After thirty years of service here at XYZ Inc. as Vice President, it is hard to believe that you have reached retirement.	在 XYZ 公司擔任三十年的副董事長一職，很難相信您已經要退休了。
We have been informed of your retirement and would like to congratulate you on your long career.	得知您即將退休，恭喜您的職涯生活圓滿結束。

表達感謝 & 讚揚對方

I appreciate your valuable contributions towards ABC Inc. wholeheartedly.	我由衷感激您對 ABC 企業的奉獻。
I have always noticed your dedication, punctuality and sincerity towards work.	我總是注意到您對工作的認真、準確以及努力。
You have always shown such competence and efficiency.	您總是那麼稱職且有效率。

You have always worked hard, fulfilled your responsibilities like a boss and done duties as an employee in your entire career.

您總是認真工作，像老闆一樣盡責，像員工一樣努力完成工作。

During your tenure here at Toyota Inc., you have generously contributed your talents and abilities.

您在豐田公司的工作期間，慷慨貢獻了您的天分與能力。

We are grateful for all your dedications and are left with no words for appreciation.

我們很感激您的付出，一切盡在不言中。

Your efforts have helped this company grow to the position it enjoys today.

您的努力使得公司擁有現在的地位。

Ever since you worked at our office, the company has grown so much in terms of scale and profitability.

自從您任職於本公司之後，公司的規模以及利潤都成長許多。

Because of you, the company has become one of the best in the business.

因為您的關係，公司成為同行間的佼佼者。

I am writing to you this letter because I wanted to thank you personally for your valuable contributions to our department.

謹以此信感謝您過去對本部門的貢獻。

You have brought fame and honor to this company.

您為公司帶來名利。

離職相關通知

I have issued a notice to Human Resources Department to prepare your documentation.

我已通知人事部門準備您的文件。

You will get all your funds and dues cleared from Account Department a day before your official farewell.

在歡送會前一天，您會從會計部拿到您的款項。

It is tough for me to see you going today, but still, congratulations on your retirement.	對我來說，看您離開很不好受，但還是要恭喜您退休。
It is hard for us to let go of an employee like you.	真的很不願意讓像您這樣的員工離開。
My eyes are wet at this moment when you are leaving us.	我的眼眶充滿淚水，因為此刻您要離我們而去。
However, I know that you have your own reasons for leaving the company.	然而，我們明白您有必須離開的理由。
If I can be of assistance, just give me a call or simply drop by at my office.	如果需要幫忙，請來電聯絡，或隨時來辦公室找我。
Mr. Joe, as you begin your retirement, we hope that you are able to fully enjoy the leisure you have earned.	喬先生，我們祝您能好好享受退休後的休閒生活。
I hope you have had a great tenure with our firm.	希望您在本公司任職愉快。
We hope you have enjoyed being here.	我們希望您很享受在這裡的日子。
Wish you a happy life ahead.	祝您往後的日子開心愉悅。
Wish you good health.	祝您身體健康。
I wish you all the best life for your retirement.	祝福您的退休生活順利。
We shall miss your visit to our company.	我們會想念您的來訪。

1-12 裁員通知

通知裁員時必須交代以下資訊：一、裁員職務；二、生效時間；三、裁員為暫時性還是永久性；四、解釋原因；以及五、後續聯絡人的資料。

From:	Fred Miller
To:	Kevin Smith
Subject:	Realigning Resources and Reducing Costs

Dear Mr. Smith:

In response to the global economic downturn, the company management has come up with the decision to **adjust**[1] company's cost structure by **realigning**[2] resources. We had been hoping that during this difficult period of **reorganization**[3], we could keep all of our employees with the company. Unfortunately, this is not the case.

It is with **regret**[4], therefore, that we must inform you that we will be unable to **utilize**[5] your services after June 30, 2018. The **layoff**[6] is **permanent**[7], and your **severance**[8] package will be paid after **submitting**[9] your documents. We have been pleased with the qualities you have exhibited during your tenure of employment with us. Thank you for your continued hard work and contributions you had brought to the company.

Please accept our best wishes for your future. If there's any question about the above-mentioned issue, please contact the Human Resources Department.

Very truly yours,

Fred Miller

親愛的史密斯先生：

因應全球經濟低迷，本公司管理階層決定調整公司的成本結構，並進行資源重整。儘管我們希望能在公司重組期間，留住所有員工，但不幸的是，這無法做到。

很抱歉，我們必須通知您：二〇一八年六月三十日後我們無法繼續讓您服務。此次裁員是永久性的，您的遣散費將於您遞交文件後發放。我們很滿意您在本公司的服務品質，也非常感謝您對公司的支持與付出。

祝您有個美好的未來，若有任何相關問題，請洽人資部門。

弗瑞德・米勒 敬上

 E-mail 關鍵字一眼就通

1 **adjust** 動 改變
2 **realign** 動 重新編制
3 **reorganization** 名 改組
4 **regret** 名 遺憾
5 **utilize** 動 利用
6 **layoff** 名 裁員
7 **permanent** 形 永久的
8 **severance** 名 切斷
9 **submit** 動 提交

 複製 / 貼上萬用句 Copy & Paste

公司重新整編

As you may know, recent changes in the economy have forced us to make some difficult decisions here at TCT Company.	如您所知，因為經濟上的波動，導致 TCT 公司必須做些困難的決定。
Due to the loss in business, we regret to inform you that we are laying off some employees for the winter and spring.	由於經營上的虧損，我們很遺憾地通知您：本公司必須在冬天和春天解僱員工。
As you have been informed, the company is reducing its workforce due to the need to reduce operating costs.	如您所知，由於公司必須減少開銷，目前正在進行人力縮編。
Our sales have dropped 40 percent in the last six months.	我方過去六個月的銷售量已降低了百分之四十。

🖱	I regret having to tell you that our services will have to be terminated.	很遺憾地告訴您，公司即將停止營運。
🖱	In order for the company to succeed in the future, we must streamline our organization today.	為了日後的成功，我們必須重整公司結構，進行縮編。
🖱	We have determined that layoffs at ABC Company are necessary.	我們已做了決斷，認為 ABC 公司有必要裁員。
🖱	The decision to eliminate jobs is a very difficult one.	裁員是一項非常困難的決策。
🖱	We regret to take this action.	我們很遺憾必須採取此行動。

通知被裁員

🖱	I write to confirm that your position has been eliminated.	此函通知：您的職務已被終止。
🖱	This is to inform you that we are eliminating your position effective immediately.	此函通知：您的職位將立即被取消。
🖱	We expect the termination of your employment to be permanent.	您的職務解除為永久性的。
🖱	I am afraid that I have to tell you bad news.	我恐怕要告訴您一個壞消息。
🖱	We will be sorry to see you go.	很抱歉，我們必須解僱您。
🖱	It is with regret that I inform you that we are eliminating your position and terminating your employment, effective as of June 30.	很遺憾要通知您：我們必須終止您的職務，此決策將自六月三十日生效。

| Due to the change of company's policy and direction this year, we are sorry your working ability does not meet our requirements any more. | 因為今年公司的政策和方向改變，很抱歉您的工作能力將不符合公司需求。 |
| It is unfortunate for our company that, due to economic constraints, we must reorganize our organization and lose valued employees. | 很不幸地，因為經濟上的限制，我們必須重組公司，並解僱部分有價值的員工。 |

裁員的補償

Thank you for your contributions for the company.	感謝您對公司的貢獻。
Thank you for your hard work and dedication at ABC Company.	謝謝您努力為 ABC 公司工作。
We have been pleased with the work that you've accomplished during your employment here.	對於您任職期間所完成的工作，我們感到很滿意。
You are entitled to two week's severance pay, which will be paid in full on your next paycheck.	您可以在下次的薪水中多領到兩週的資遣費。
We will provide severance pay in the amount of $20,000.	我們會提供兩萬元的資遣費。
You will receive your severance pay and final paycheck on July 5.	您會在七月五日同時收到資遣費和最後一份薪資。
You will get all legal welfare in accordance with the Labor Law.	您會得到勞基法所規定的合理福利。

裁員相關提醒

On your last day of work, June 30, please be sure to return your keys and any other items belonging to the company.

在您最後的工作日（六月三十號）那天，請歸還公司鑰匙以及物品。

Please arrange for the return of any company property in your possession.

請事前安排好，歸還您所使用的公司物品。

Related information regarding this layoff will be provided to you later.

有關此次的裁員資訊，會在之後寄給您。

Enclosed you will find information regarding your transition and your opportunity to continue healthcare coverage.

請您參考附件裡關於健保支付的說明，以及這段過渡期所需要的協助。

信件結語：聯繫 & 祝福

Please apply again for our next summer season.

請於明年夏季再應徵。

If you have any additional questions, please do not hesitate to call me.

如果您有任何疑問，請不用客氣，隨時電話聯繫我。

Should you have any questions, please contact Paul Lai: 04-2227-3358.

如有任何疑問，請與賴保羅聯絡，他的電話是：04-2227-3358。

Please accept our best wishes for you in the future!

請接受我們對您未來最衷心的祝福！

We wish you all the best in your future endeavors.

祝福您在新的工作崗位上順利。

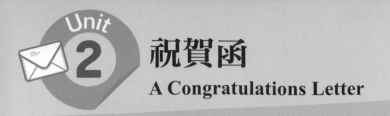

2-01 生日祝福

　　生日對於一個人來說，是一年中最特別的一天。無論是面對同事還是客戶，最好都在這一天獻上祝福，表示關心。若對方有生日宴會，你卻無法參加時，也最好提前回覆，於信中委婉告知原因，並給予祝福。

From:	Polly
To:	Charlotte
Subject:	Happy Birthday

Dear Charlotte:

It came to my mind that your birthday is just around the corner. I am sure you must have planned a lot for it - a big rocking dancing party or something. Have you heard about the saying: "Growing old is **inevitable**[1], while growing up is **optional**[2]."? I feel this holds true for you. Each time I meet you, I see a child somewhere in you. You are always so **lively**[3]. I really need to learn from you.

I pray to God for a long, healthy and **rocking**[4] life for you. May He shower all His **grace**[5] on you and your family!

Enjoy yourself on your birthday and every day of the year, too.

Take care!

Love,

Polly

親愛的夏綠蒂：

我想起您的生日快到了，相信您一定已經規劃了許多活動，像是舉辦一場舞會之類的。不知道您是否聽過一句話：「年齡的增長雖不可避免，但成長與否，卻可以由我們自己決定。」我覺得這句話很適合您。每次見到您，您總是生氣勃勃、充滿活力，我需要向您學習。

希望上帝賜予您健康精彩的人生，也希望祂保佑您與家人。

祝您生日那天以及往後的每一天都能愉快地度過。

保重！

愛您的波莉 敬上

 E-mail 關鍵字一眼就通

1 **inevitable** 形 必然的　　2 **optional** 形 可選擇的　　3 **lively** 形 精力充沛的

4 **rocking** 形 精彩的　　5 **grace** 名 （神的）恩典

 複製 / 貼上萬用句 Copy & Paste

信件開頭

Next week will be your birthday.	下週是您的生日。
It is your 30th birthday.	您三十歲的生日到了。
So your birthday is coming.	您的生日即將到來。
Congratulate you on your 40th birthday.	祝賀您四十歲生日。
Let me wish you happy birthday in advance.	請讓我提早祝您生日快樂。
Your birthday is coming next week. Enjoy it to the fullest!	下週就是你的生日了，好好慶祝吧！

無法參加生日宴會

I am really sorry to inform you that I cannot attend your birthday party.	很抱歉，我無法參加您的生日宴會。
I am very sorry for my absence from your birthday.	很抱歉，您生日那天我無法出席。
I will miss all the fun that we would have had there, and I will miss you.	雖然我無法與您同樂，但我會想念您。
I wish I could personally come to share the joy on this milestone in your life, but I happen to be so far away from you physically at the moment.	您那天將邁入新的里程碑，真心希望能到場分享您的喜悅。但不巧的是，我那天剛好身在遠方。
I can't congratulate you in person.	我無法當面祝賀。

送上生日祝福

I send this mail to offer you my sincere congratulations and best wishes for many happy returns of your birthday.	在此獻上我衷心的祝福，也祝您生日那天能開心地度過。
I heartily congratulate you for your birthday.	衷心祝福您生日快樂。
May you enjoy good health and long life!	祝您健康長壽！
Accept my wishes for many happy returns on your birthday!	祝您生日那天過得愉快！
Blowing out another candle should mean you have lived another year of joy.	吹熄蠟燭代表您又過了美好的一年。
Keep this spirit up and have a delighted birthday!	繼續保持下去，祝您生日快樂！
Happy birthday (to you)!	（祝您）生日快樂！

🖱 Happy birthday with love!		祝福您生日快樂！
🖱 Hope your 40th birthday is happy!		祝福您四十歲生日快樂！
🖱 Happy birthday! Hope this is the beginning of another great and powerful year to you!		生日快樂！祝您接下來的這一年也能過得很棒，並充滿活力。

信件結語

🖱 I will soon come to visit you.		我近期會去拜訪您。
🖱 Hope this is a special birthday for you!		祝您有個難忘的生日！
🖱 Wish you have a great birthday and a year filled with happiness!		祝您有個很棒的生日，一整年都快樂！
🖱 May your future life full of prosperity and happiness!		祝您未來充滿快樂和成就！

實力大補帖 Let's learn more!

活字典 ▲ 單字 / 片語集中站

📍 布置生日派對

balloon 氣球	birthday hat 生日帽
candle 蠟燭	catering 外燴餐點
confetti 五彩碎紙	dance floor 舞池
decoration 布置	refreshments 點心
tissue pom poms 紙球花	

From:	Lillian Chen
To:	Marvin Johnson
Subject:	Congratulations on the Marriage

Dear Marvin:

I was really full of joy to hear about your big day. Next week your **engagement**[1] is there and next month you are going to **get married**[2]. I am giving my best wishes to you for both **occasions**[3].

I am feeling very happy that you both are going to step in your new life. My best wishes are always with you. May God give you all the happiness in the world! However, I am feeling really sorry to tell you that I cannot come for your **wedding**[4] and even for the engagement. Next week I am going to Europe for an office meeting. It is really an important meeting and cannot be **cancelled**[5]. I tried a lot to come but it is not **possible**[6] for me.

Still my wishes are with you.

Best regards,

Lillian Chen

親愛的馬文：

很高興聽到您要結婚的消息。聽聞您下週就要訂婚，下個月即步入禮堂，我由衷祝福這兩件人生大事！

很高興您與夫人即將攜手邁向人生的另一個階段，真心地祝福您，願主賜予您喜樂！但很抱歉的是，我無法參加您的訂婚以及婚禮。下週我必須到歐洲參加會議，而且是場極為重要、無法取消的會議。雖然我試著安排，但實在是不克前往。

無論如何，您仍擁有我最衷心的祝福。

莉莉安‧陳 敬上

key words ▲ E-mail 關鍵字一眼就通

1 engagement 名 訂婚　　2 get married 片 結婚　　3 occasion 名 場合

4 wedding 名 婚禮　　5 cancel 動 取消　　6 possible 形 可能的

✈ 複製 / 貼上萬用句 Copy & Paste

信件開頭

🖱 Hi! How have you been recently?	嗨！您們最近好嗎？
🖱 Congratulations on your marriage!	恭喜您結婚！
🖱 What a great news!	真是條大新聞！
🖱 I am so happy that you have finally decided to settle down in life.	很高興您終於決定要安定下來了。
🖱 I am truly happy for you after hearing the news.	得知此消息後，我真是為您感到開心。
🖱 I am really happy for you!	我真替你感到高興！

分享心情 & 其他資訊

🖱 Let me tell you by my experience that marriage is one of the most wonderful things to happen to a person.	讓我與您分享，婚姻是人生中最美好的事之一。
🖱 A relationship, a sharing, a bond and loads of love and happiness – that is what marriage is.	一段愛與快樂的關係、分享、牽絆與責任，這就是婚姻。
🖱 Needless to say, you can always count on me for any help needed during the wedding time.	不用說，婚禮需要幫忙時，您可以隨時找我。

I am really sorry for my absence on that day.	很抱歉，我那天會缺席。

給予結婚祝福

Please convey our heartiest congratulations to Rita and your family as well.	請接受我對芮塔與您家人的誠心祝福。
Take care and stay in touch.	保重並保持聯絡。
Please accept my heartfelt congratulations.	請接受我衷心的祝福。
To the Bride and Groom, I wish you love and happiness!	祝新娘新郎永浴愛河！
Please accept my congratulations and my best wishes to you for a marriage filled with all the good things in life.	請接受我的祝賀，祝您婚後生活事事如意。
You two have my earnest and warmest congratulations on this event.	獻上我對此盛事最衷心的祝福。
You two make a perfect couple.	您倆是天生一對。
Good wishes for a long and happy life!	祝您白頭偕老！
Best wishes for many years of happiness for you two.	祝百年好合。
I extend my best wishes and congratulations to you.	我表達對您的祝賀。
Please accept my best wishes and congratulations.	請接受我的祝賀。
May you have a perfectly happy marriage!	祝福您婚姻美滿！

🖱 **We would like to extend our best wishes for the future.**　對於您的未來，我們在此獻上最衷心的祝福。

🖱 **Mr. Martin joins me in sending kindest regards and best wishes to you all.**　馬丁先生和我一起問候各位，並致以美好的祝福。

實力大補帖 Let's learn more!

活字典 單字 / 片語集中站

📍 與結婚典禮相關的人們

bride 新娘	groom 新郎
bridesmaid 伴娘	best man 伴郎
flower girl 女花童	ring bearer 男花童
guest 嘉賓	emcee 司儀；主持人
photographer 攝影師	matchmaker 媒人

深入學 資訊深度追蹤

📍 各國婚禮的小知識

01 法國：以白色表示「純潔無瑕」的概念，從花束到禮服，都會用白色裝飾。

02 希臘：會於手套中放一些糖，表示「婚姻生活會很甜蜜」。

03 德國：應邀前來的賓客會帶著不用的舊碗盤，於門口打破，表示新人擺脫了舊日的各種煩惱。

04 英國：傳統英國婚禮中，新娘必須穿戴「舊的、新的、借來的、以及藍色的物品」，以表示婚姻生活的美滿。

在比較重要的節日，往往會寄送信件給人，送上祝福。一般而言，幾個比較重要的節日皆會如此。例如：台灣有農曆春節、中秋節；美國則是元旦、聖誕節、感恩節。其他節日則視個人情況而定，例如本章收錄的情人節，近來也成為慶祝的一大節日。

近來，祝賀信的內容愈來愈生動活潑，也更具創意。寄件者可依據與對方的熟識程度，決定信件的內容走向。不過大體說來，只要記得在這些重要節日給予祝福，都會讓人感到非常窩心。

From:	Linda Yang
To:	sammy@mail.net
Subject:	Happy New Year!

Dear friend:

It's been an eventful year.

2017 was simply **incredible**[1] and **memorable**[2] for all of us, all that craziness **amidst**[3] the hard work. Still, we **pulled off**[4] an amazing job together.

I wish your New Year **countdown**[5] to 2018 be filled with kisses, happy roar and cheerful **toasts**[6]! Wherever you are, have fun!

Also, I wish you all and your families good health, love and happiness in 2018!

Cheers,

Linda Yang

親愛的朋友：

今年真是個多事之年。

二○一七年對我們來說實在太令人難忘，忙碌的工作也讓人五味雜陳。儘管如此，我們還是成功地完成了工作。

希望您二〇一八年的跨年倒數充滿香吻、歡樂的喧鬧聲以及愉快的敬酒！不論您在哪，祝您玩得愉快！

同時也祝福您的家人在二〇一八年身體健康、平安快樂！

楊琳達 上

From:	Luna Smith
To:	all@mail.net
Subject:	Happy New Year!

To all friends from Apple Company:

Thanks for the great memories in 2017!

Here I want to wish you a blessed and a holy Christmas, and also Happy New Year in 2018!

Take care and keep in touch!

Best,

Luna Smith

給所有蘋果公司的朋友：

感謝二〇一七年與您度過的美好回憶！

祝福您聖誕佳節以及新年愉快！

保重，隨時保持聯絡！

露娜・史密斯 敬上

From:	Doris Chou
To:	all@mail.net
Subject:	Valentine's Greetings!

Dear all:

As the **world**[1] celebrates this day of **hearts**[2], take a **moment**[3] and give a kiss, even a **hug**[4] to everyone special in your life.

I hope you could **blow a kiss**[5] my way. =)

Happy **Valentine's**[6] Day!

Doris Chou

親愛的各位：

當全世界都在慶祝這個愛的節日時，記得給身邊那位特別的人一個香吻及擁抱。

也希望您給我一個飛吻（笑）。

情人節快樂！

朵莉絲・周 上

 E-mail 關鍵字一眼就通

1 world 名 世界　　　**2** heart 名 感情　　　**3** moment 名 片刻

4 hug 名 擁抱　　　**5** blow a kiss 片 飛吻　　　**6** valentine 名 情人

Part
1
入門篇

Part
2
求職篇

Part
3
客戶往來篇

Part
4
人際互動篇

From: Nancy Brown

To: all@mail.net

Subject: Merry Christmas!

Dear customers:

The holiday season is a **wonderful**[1] time for us to **remember**[2] the friends and **customers**[3] who **sustain**[4] our business and make our jobs a **pleasure**[5] all year long. Our business would not be possible without your continued support.

So we'd like to take this moment to say thank you and send our best wishes to you and your families. May your new year be filled with **success**[6] and happiness!

Happy Holidays!

Nancy Brown

ABC Company

親愛的客戶：

這個節日讓我們想到長年支持我們的朋友與客戶。沒有您的支持就不會有現在的我們。

我們希望趁此機會向您致謝，並向您及家人獻上我們最深的祝福。希望您來年事事成功、順心快樂。

祝 佳節愉快！

ABC 企業 南西‧布朗 敬上

key words ▲ E-mail 關鍵字一眼就通

1 wonderful 形 極好的　　**2** remember 動 記得　　**3** customer 名 顧客

4 sustain 動 支持　　**5** pleasure 名 愉快　　**6** success 名 成功

From:	Rita Smith
To:	Mr. Washington
Subject:	Merry Christmas!

Dear Mr. Washington:

As the holiday season approaches, we'd like to take this opportunity to thank you for your continued business. It is business associates like you who make our jobs a pleasure and keep our business successful.

May your holiday season and the New Year be filled with much joy and happiness. We look forward to working with you in the coming year and hope our business relationship continues for many years to come.

Happy Holidays!

Rita Smith

華盛頓先生您好：

佳節將近，因此想趁此機會感謝您一直以來對我們的支持。因為有您這樣的合作夥伴在，才讓我們工作起來既愉快又順利。

在此祝您有個美好的聖誕假期及新年，盼望來年也有機會繼續合作，也期望我們的合作關係能長久。

祝 佳節愉快！

芮塔‧史密斯 敬上

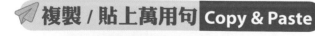 複製 / 貼上萬用句 Copy & Paste

聖誕節暨新年祝福

 Best wishes for a merry Christmas! 聖誕節快樂！

	Wish all the blessings of a beautiful Christmas season.	願您擁有聖誕節的祝福。
	As the Christmas season approaches, may you and your family all the joy and happiness that you deserve!	耶誕節將近，祝您和家人得到應有的喜悅和祝福！
	Best wishes for Christmas and the Chinese New Year!	恭賀聖誕節以及新年快樂！
	Best wishes for Christmas and the New Year!	祝聖誕暨新年快樂！
	We wish you and your staff a Merry Christmas and a Happy New Year.	謹祝福您與員工聖誕暨新年快樂。
	Best wishes for the New Year!	新年快樂！
	The happiest New Year to you and your family.	祝您和家人新年愉快。
	May the New Year find you in the enjoyment of health and happiness!	祝您新年愉快平安！
	I wish you a happy New Year.	祝您新年快樂。
	Happy Holidays!	佳節愉快！

送上其他祝福

	We are very pleased to send you and your colleagues our best wishes.	我們很高興能向您和您的同事致上最美好的祝福。
	We wish you continued success in the coming year, and all the best to your business.	祝貴公司來年業務蒸蒸日上，也祝您事業有成。
	May all your wishes fulfilled!	希望您的願望都能實現！
	May you achieve your goals!	祝您順利達成目標！

We wish you a Happy New Year, a year big with achievement, a year mellow with happiness and content.	希望您來年有成就且快樂滿足。
May you and your family health, happiness, peace and prosperity this year!	祝您與家人新的一年裡能健康愉快、平安興盛！
Allow me to congratulate your family on this special day.	在這個特別的日子裡，謹向您的家人表示祝賀。
May fortune smile upon you and favor you with many blessings.	希望幸運降臨並賜福於您。
May the coming year bring you happiness in fullest measure.	希望您未來的一年歡樂滿溢。
May the New Year bring you health, happiness and all other good things!	祝您來年健康快樂、事事順心！
May the New Year be a good year to you and yours - full of health and happiness!	祝您與家人來年健康快樂！
May each of the three hundred and sixty-five days of the New Year be the happy one for you!	祝您未來三百六十五天每天都快樂。
Warm greetings and best wishes for the happiness in the coming year.	衷心祝您明年快樂。
May peace and happiness be yours in the New Year!	祝您在新的一年中平安快樂！
Wish you health and happiness in the year to come!	祝您來年健康快樂！

2-04 生產祝賀

在「生產祝賀」的信件中，必須充分表達自己的心情和祝福，因此內文可以多敘述一些自己的想法，但必須禮貌得體。

From:	Anna Lee
To:	Hans
Subject:	Congratulations on the Birth

Dear Hans:

I have just heard the wonderful news about the birth of your son. Congratulations!

A child is the most **precious**[1] gift you have in your life. Please enjoy every moment spent with the baby. We hope that the baby will grow to be a good **citizen**[2], and bring **honor**[3] and **glory**[4] to your family.

Again, I am very happy for you and your wife. Hope your wife will **recover**[5] soon after **labor**[6].

Best regards,

Anna Lee

漢斯您好：

我剛得知令郎誕生的好消息，恭喜您！

孩子是人生中最珍貴的禮物，請好好享受與孩子相處的每個時光。希望您的孩子能健康成長，成為一個好公民，並為您的家族帶來榮耀。

我要再次表示，我為您與尊夫人感到開心，望尊夫人的產後復原情形良好。

李安娜 敬上

1 precious 形 珍愛的　　**2** citizen 名 市民　　**3** honor 名 榮譽

4 glory 名 光榮　　**5** recover 動 恢復　　**6** labor 名 分娩

複製 / 貼上萬用句 Copy & Paste

聽聞消息 & 祝賀

Congratulations to you and your growing family!	恭喜您的家族又增添新成員了！
It's so great to know that the baby has arrived healthily.	很高興得知您的孩子健康出生。
How wonderful to hear that you had a baby boy!	很高興聽到您喜獲麟兒！
Please forgive me for writing so late to congratulate you on the birth of your son.	很抱歉這麼晚才寫信祝賀令郎的誕生。
I came to know about this news two days ago.	我於兩天前得知此消息。
I came to know that the mother and the child are both well.	我獲知母子均安的消息。
It is really good to know that your wife and your son are safe and sound.	很高興得知您的夫人與孩子都平安。
I heard the good news and I was very delighted by it.	我得知消息後非常開心。
It is really great of having new addition in your family.	貴府喜獲麟兒真的是件很棒的事。
Let me congratulate you two on having a newborn baby!	恭喜二位喜獲新生兒！

We are delighted to hear that your wife had given birth to a son.	恭喜您與尊夫人喜獲麟兒。	Part **1** 入門篇
We are so excited to hear your great news.	我們很興奮得知這麼令人開心的消息。	Part **2** 求職篇
It is so great to hear that you have added another member to your family.	很高興得知您的家族多了一位新成員。	
What wonderful news, you and your wife must be very happy.	真是個天大的好消息,您與尊夫人一定非常開心。	Part **3** 客戶往來篇
Your wife and family must be very happy now.	尊夫人與您的家人現在一定非常高興吧!	

為新生兒送上祝福

All the best to your little baby boy.	祝福令郎。
Best wishes for the arrival of your daughter.	祝福令嬡的誕生。
She must be a lovely baby girl.	她一定是個可愛的女寶寶。
I am sure the baby will be beautiful and gifted just like her mother.	我相信寶寶將來會和她母親一樣,既美麗又有才華。
Congratulations on both of you, and wish the new arrival happiness and health.	謹此祝賀兩位,並祝新生兒健康快樂。
Please accept our warmest congratulations to you.	請接受我們最誠摯的祝福。
I want to convey my hearty congratulations to you regarding your new baby.	我想向您的孩子獻上我衷心的祝福。
All of us here at the Human Resources Department send our congratulations on the birth of your baby.	人資部門的所有同仁為您孩子的出生致上祝賀。

| I pray to God that he blesses your child and gives him a good and a bright future. | 我向上帝祈禱，希望您的孩子有個美好的未來。 |

關懷與提醒

Please take care of your baby and your wife.	請照顧好您的寶寶和太太。
You need to take care of their health for a couple of months.	這幾個月，您要照顧他們的健康。
But you still need to take care of them.	但您仍須照料他們。
Being a father is a lifetime dream, but at the same time, there are lots of responsibilities.	成為父親是夢寐以求的事，但同時也帶來許多責任。
Give proper medicines and vaccinations to your child as the newborns are at a threat of getting many diseases.	為避免染上疾病，記得要定期帶新生兒去施打預防針。
But now you have many challenges to face for the upbringing of the child.	不過，您現在也要開始面對許多育兒的問題。
There are various difficulties in the upbringing of a baby.	養育孩子將會有許多難題。
New responsibilities are with you now.	您即將承擔新的責任。
You should spend time with your family now.	您現在應該多陪陪家人。
You should cut down some pressure of your work to look after your family.	您應該減少工作方面的壓力，以便多照顧家人。
You need to plan about the future for your children.	您應該開始計畫孩子們的未來。

Thus you should understand things and try to bring a better future for your baby.

因此您應該要開始明白事理，並給您的孩子一個美好的未來。

前往拜訪

Will you send me a photo of the baby?

您可否寄張寶寶的照片給我呢？

We will be happy to come and visit your families.

我們想要拜訪您的家人。

I cannot wait to see your new baby.

我等不及要看寶寶了。

When the mother is well enough and the baby is a bit older, we would like to call on you.

等媽媽的身體狀況好一點，孩子也大一點後，我們就會去拜訪您。

信件結語

This is a new beginning of your life.

這會讓您的人生展開新的一頁。

Please enjoy everything while you can.

請好好享受每一刻。

We are sure you will be a good father.

我們相信您會是個好父親。

We are sure your son will make you proud one day.

我們相信，將來令郎會讓您感到非常驕傲。

I wish God blesses your child.

祈禱上帝保佑您的孩子。

We wish your families all the happiness!

我們希望您的家庭幸福快樂！

I wish you all happiness with your new little son and your families.

希望令郎和您的家人都無比快樂。

Wish you and your baby good health, love, and happiness forever.

祝福您與孩子常保健康、幸福、快樂。

From:	Billy Jones
To:	Charlotte
Subject:	Congratulations on the Promotion

Dear Charlotte:

Congratulations on your new position as **Marketing**[1] Manager! I was not **surprised**[2] by the news, because you have always showed your **enthusiasm**[3] when you work in the past years. As people say, "Hard work **pays off**[4]."

I would like to express my **hearty**[5] congratulations on this good news. We are sure that you will make great success in your position.

Best regards,

Billy Jones

親愛的夏綠蒂：

恭喜您升任行銷主管！聽到這個消息我一點也不驚訝，因為過去幾年，您總是懷抱熱忱，並努力工作著。俗話說得好：「有努力便有回報。」

在此獻上我誠心的祝賀，我們相信您會在這個職位上發光發亮。

比利·瓊斯 敬上

 E-mail 關鍵字一眼就通

1 marketing 名 行銷　　**2** surprised 形 驚訝的　　**3** enthusiasm 名 熱情

4 pay off 片 成功　　**5** hearty 形 熱誠的

複製 / 貼上萬用句 Copy & Paste

升遷消息 & 祝賀

I have just learned that you have been appointed as Chairman for the Asia Area.	我剛得知您被任命為亞洲地區的總裁。
I am very delighted to hear that you have been appointed as General Manager.	很高興您被指派為總經理。
It was a great pleasure to hear of your appointment as President.	很開心得知您被指派為董事長。
May we congratulate you on your promotion to Marketing Manager!	恭喜您升任行銷部經理！
Please accept our congratulations on your promotion to Manager of the Marketing Department.	恭喜您當上行銷部主管。
I am very happy to hear about your promotion.	我很開心聽到您升遷的消息。
Congratulations on your promotion!	恭喜您升遷！
I am so happy to learn about your promotion.	很高興得知您升遷的消息。
We'd like to send you our warmest congratulations on the occasion of your new appointment as Vice President.	我們誠摯地恭賀您就任副董事長。
Congratulations on your recent promotion to Executive.	恭喜您晉升為執行長。
It was my pleasure to learn that you are appointed as Marketing Manager of your company.	很高興您晉升為貴公司的行銷經理。
Congratulations on becoming an important member of your company.	恭喜您成為貴公司的重要成員。

Allow us to extend our warmest congratulations on your promotion.	請讓我們由衷地祝賀您獲得升職。
May we extend to you the warmest greetings on your promotion.	請容我們獻上恭賀，恭喜您升遷。
All the employees want to congratulate you on your promotion.	全體同仁恭賀您榮升。
I was truly happy for you at the news that your husband was promoted to Chief of Marketing Department of XYZ Company.	聽聞您的先生晉升為 XYZ 公司的行銷主管，我由衷地替您感到開心。
Please convey to him my heartiest congratulations.	請轉達我對他真誠的祝賀。

進一步給予肯定

You are a hard worker.	您的工作態度很認真。
You have worked so hard these years.	您過去幾年如此認真工作。
We all know you have worked hard for many years for the company.	我們都知道您過去幾年為公司努力工作。
You always showed your dedication while working.	您過去工作總是很盡責。
I know how hard you have worked to earn the recognition you presently enjoy at ABC Inc.	我知道您是多麼辛苦的工作，才能贏得在 ABC 企業的地位。
Working with you in the past seven years has shown me how capable and enthusiastic you are.	在過去七年與您的共事中，一直覺得您才能出眾，對工作也懷有熱忱。
Looking back on your activities, we all know that your experiences are the qualities that this position requires.	回顧您以往的工作，我們都清楚您的經驗很符合這個職位的要求。

Having worked with you so long, I can confidently say that you are the right person in the right position.	與您共事這麼久，我敢說您是此職位的最佳人選。
As your coworker, I am proud of your hard-working attitude and promotion.	身為同事，我以您的工作態度及升遷為榮。
Your success is certainly the fruit of your tireless effort.	您的成功是您不屈不撓努力的結果。
You are one of the most capable women I have ever met.	您是我所認識的女性中，最有能力的人之一。
I feel that your company is very wise in making this decision.	我認為貴公司做了一項非常明智的決定。
You really deserve it!	這是您應得的！

恭賀就職

Congratulations with your new job.	恭喜您找到新工作。
I am very happy to hear that you have secured a job in Apple as a software engineer.	很高興您獲得蘋果電腦的聘任，擔任軟體工程師一職。
I was thrilled to hear about your new job in ABC Company.	我很開心聽到您在 ABC 公司找到新工作。
Congratulations my dear friend and keep it up.	恭喜您，我親愛的朋友，請繼續努力。
Please accept my best wishes for every success in your new position.	請接受我衷心的祝福，並祝您在新職務上能夠成功。
I know it's been a long search to find the right position, but it seems like this is going to be a good match for your skills and experience.	我知道您已求職許久，看來這是一份很符合您能力以及經驗的工作。

Your enthusiasm and your coordination abilities are certain the qualities to support you in the new position.	您的熱忱和協調能力都將是支持您任居此職位的特質。
I am sure your professional experience has prepared you to take such challenges.	相信您的專業已經讓您準備好接受挑戰了。
It's a great achievement to get a job in such a huge multinational company.	能夠在如此大型的跨國企業工作，真是莫大的成就。
Congratulations on the new position. I believe you will juggle the responsibilities of family and work just fine!	恭喜您任職新職位，相信您一定能在家庭與工作間取得最佳平衡！
You will meet many intelligent and professional personals in the office.	您會在辦公室遇到許多聰明的專業人士。
Now you must work hard to stay among them, and learn more and more things from them.	現在，您必須在他們之間認真學習許多事。
I know our colleagues will enjoy working with you.	我相信同事們都會很高興與您一起工作。
We are sure you will make great achievement for the organization through this opportunity.	我們相信您會以此契機，為公司創造更多成就。
I am confident that you are more than equal to the difficult task that awaits you.	我相信以您的能力，處理難題是綽綽有餘的。

歡迎新就職的同事

Welcome to our office.	歡迎來到我們辦公室。
Welcome to our company.	歡迎來到我們公司。

🖱 We are so happy to have a new colleague.	我們很開心有新同事加入。	
🖱 We are most fortunate to have you.	很榮幸有您的加入。	
🖱 Great to hear that you will join our team.	聽聞您即將成為我們的一員，真是感到高興。	
🖱 Please let us know what we can do to help you get settled.	若需要幫忙，請隨時告知。	
🖱 We look forward to your continued success.	我們期盼您越來越成功。	

信件結語

🖱 It's our sincerest wish that under your democratic leadership, your company will continue to prosper.	祝福貴公司在您的開明領導下，業務蒸蒸日上。
🖱 I expect this advanced position will challenge and teach you even more about strategies in marketing.	我期待這個高階職位將更有挑戰性，並能讓你學到更多的行銷策略。
🖱 I wish you every success in managing the affairs of the organization.	祝福您在管理公司事務上能夠成功。
🖱 We feel the appointment is timely as well as it is well deserved.	我們覺得這項指派來得正是時候，也恰如其分。
🖱 You will definitely make greater success in the near future.	您必定會在不久的將來獲得更大的成功。
🖱 Your future is brilliant!	您的未來一片光明！
🖱 Good luck to you!	祝您好運！
🖱 I'm looking forward to hearing all about your new job soon!	希望近日能夠聽到關於您新工作的消息！
🖱 Best luck in your new job.	祝福您在新工作上表現出色。

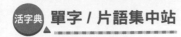
活字典 單字 / 片語集中站

📍 公司各部門

Head Office 總公司	Branch Office 分公司
Business Office 營業部	Secretarial Pool 祕書室
Export Dept. 出口部	Import Dept. 進口部
Sales Dept. 業務部	Personnel Dept. 人事部
HR Dept. 人資部	Public Relations Dept. 公關部
Marketing Dept. 行銷部	Advertising Dept. 廣告部

📍 公司內的職稱

Chairman 總裁	Vice Chairman 副總裁
President 董事長	Vice President 副董事長
Executive 執行長	General Manager 總經理
Consultant 顧問	Manager 經理
Assistant Manager 副理	Junior Manager 襄理
Director 協理	Section Manager 課長
Account Manager 主任	Engineer 工程師
Supervisor 組長	Specialist 專員
Assistant 助理	Team Leader 領班

2-06 祝賀得獎

From: Chad Harris

To: Lucy Evans

Subject: Congratulations on the Award

Dear Lucy:

I read with great **delight**[1] your name **published**[2] in today's morning paper. You have won the Turing Award for 2018, one of the most famous **prizes**[3] for computer science in the United States. Your **untiring**[4] **industry**[5] and dedication to **database**[6] technology has won you this honor.

My sincerest wishes are with you. I join with your friends and admirers in offering you my heartiest congratulations. May your **unremitting**[7] endeavor in the computer science go on with **undiminished**[8] **vigor**[9]!

Yours sincerely,

Chad Harris

露西您好：

我在今日的早報上看到您的名字，很開心看到您得到美國電腦界知名的二〇一八年圖靈獎。您在電腦資料庫不懈的鑽研及努力讓您獲得此殊榮。

請讓我加入您親朋好友的行列，致上我最誠摯的祝福。希望您的努力不懈，能夠使電腦科學更茁壯！

查德‧哈里斯 謹上

key words E-mail 關鍵字一眼就通

1 delight 名 愉快

2 publish 動 刊登

3 prize 名 獎賞

4 untiring 形 不屈不撓的

5 industry 名 勤勉

6 database 名 資料庫

7 unremitting 形 不懈的　　**8** undiminished 形 未衰減的　　**9** vigor 名 精力

複製 / 貼上萬用句 Copy & Paste

It was with great pleasure that we noticed in The Washington Post that you have received the award for service to American industry.	很開心在《華盛頓郵報》上看到您得到美國工業服務獎。
I am delighted to know that you finished the project.	很開心得知您完成了那項計畫。
We are so happy you made it!	我們很高興您辦到了！
We would like to congratulate you on this well-deserved honor.	我們想要恭賀您，這次獲獎真是實至名歸。
We are writing immediately to offer our sincere congratulations.	我們立刻致函表示誠摯的祝賀。
Allow me to offer my sincerest congratulations on your success.	讓我對您的成功致上最誠摯的祝福。
Best wishes for your bright prospects.	祝福您前景一片看好。
Congratulations on winning the game.	恭喜您贏得比賽。
Congratulations on winning the prize!	恭喜您獲獎！
Congratulations on your achievement.	恭喜您達成這樣的成就。
I am writing to congratulate you on your success.	來函祝賀您成功。
Please accept our sincerest congratulations.	請接受我們誠摯的祝福。
Please accept my congratulations on your success.	請接受我對您成功的祝福。

2-07 祝賀公司成立

From:	Victor Wang
To:	Mr. Thomas
Subject:	Congratulations on Opening a New Company

Dear Mr. Thomas:

I am pleased to know that your company will be open next week. Please accept my **warmest**[1] congratulations and best wishes.

With your excellent **leadership**[2] and **exceptional**[3] **abilities**[4], I am sure the new company will have a **brilliant**[5] future and **booming**[6] business.

Best regards,

Victor Wang

湯瑪斯先生您好：

很高興知道您的公司將於下週開幕，請接受我誠摯的恭賀與祝福。

在您優越的領導下，我相信貴公司的未來將一片光明，並且生意興隆。

維克多・王 敬上

keywords ▲ E-mail 關鍵字一眼就通

1 **warm** 形 溫暖的　　2 **leadership** 名 領導力　　3 **exceptional** 形 卓越的

4 **ability** 名 能力　　5 **brilliant** 形 光輝的　　6 **booming** 形 興旺發達的

I am very happy to know that you have established a company.	很高興得知您創辦了一間公司。
We are delighted to know that your new company is starting business today.	我們很高興得知貴公司於今日開始營業。
I'd like to add my congratulations to the many you must be receiving.	我也在此加入向您祝賀的行列。
With your excellent background and achievement, I am sure that the new company will be prosperous day by day.	以您傑出的背景及成就，一定會使您的公司欣欣向榮。
As your company continues to grow and prosper, let me know if I can assist you.	在貴公司成長與發展的階段，如有可以效勞之處，煩請告知。
You can expect our full support in the years ahead.	關於貴公司今後的發展，本公司願意全力支持。
We wish you every success and happiness in your important position.	希望您在要職上獲得成功。
Best wishes for setting a new record!	預祝創造佳績！
Good luck in all that you will achieve in the future.	祝您未來一切順利。
I congratulate on your new business.	恭祝您鴻圖大展。

實力大補帖 Let's learn more!

文法王 ▲ 文法 / 句型解構

📍 sure（確定）的三種用法

1 be sure of 後面接「名詞」，表示「確信某件事」。

ex. I am sure of your success.
我確信您將成功。

2 be sure that 後面接「子句」。

ex. We are sure that you will be successful.
我們確信您將會成功。

3 be sure to 後面接「原形動詞」，表示「做某事」。

ex. Jenny is sure to write to you.
珍妮一定會寫信給你。

深入學 ▲ 資訊深度追蹤

商業書信中常見的祝賀主題有以下幾個：

1 訂 / 結婚：常見以整個部門為代表發送祝賀信。

2 新生兒：除了一般的生產祝賀以外，由於美國社會流行領養小孩，因此若有此情形，公司也會當成家庭有新生兒而表達祝賀。

3 喬遷新址：可於祝賀信中恭喜對方因業績增長而得以喬遷新址，並可請對方提供新地址，以便後續聯繫。

4 節慶的祝賀：新年與聖誕假期是西方社會的重要節慶，常會舉辦活動，因此這類的祝賀信中通常會邀請對方來參加相關活動。

3-01 邀請對方參加公司宴會

　　在「參加公司宴會」的邀請函中，記得在本文中點明「宴會的時間與地點」，並建議於文末再度誠摯地邀請收件者參加。

From:	Kelly Wang
To:	Mr. Black
Subject:	20th Anniversary Invitation

Dear Mr. Black:

To **celebrate**[1] the 50th **anniversary**[2] of Tokyo **Motors**[3] Taipei Branch, we are planning a dinner party at the Grand Hotel in Taipei from 6:00 to 10:00 p.m. on Saturday, March 20, 2018.

You are **cordially**[4] invited to the party so that we can express our sincere appreciation to you for the **generous**[5] support you have **extended**[6] to us for a long time. For your information, the party will be attended by many top **executives**[7] of leading Taiwanese and Japanese auto manufacturers. We believe this will be an excellent opportunity for you to get **acquainted**[8] with each other.

We hope that you will be able to join us in this event to meet the **senior**[9] directors of our company.

We look forward to seeing you on Saturday, March 20, 2018.

R.S.V.P. by e-mail at admin@tokyomotor.com.tw

With warmest regards,

Kelly Wang

布萊克先生您好：

為慶祝東京車業台北分公司創立五十週年，我們將於二〇一八年三月二十日（星期六），晚上六點到十點在台北圓山大飯店舉辦慶祝酒會。

為了感激您長久的合作與支持，我們誠摯地邀請您前來參加宴會。此外，還有許多台日車產業界的高級主管也會來參與，我們相信這是您拓展人脈的好機會。

期盼您能參與酒會，並認識本公司的高階主管們。

期待於三月二十日見到您。

敬請回覆至信箱：admin@tokyomotor.com.tw。

王凱莉 敬上

 E-mail 關鍵字一眼就通

1 celebrate 動 慶祝
2 anniversary 名 週年紀念
3 motor 名 汽車
4 cordially 副 誠摯地
5 generous 形 大方的
6 extend 動 給予
7 executive 名 主管
8 acquaint 動 使認識
9 senior 形 高階的

 複製 / 貼上萬用句 Copy & Paste

邀請：公司或個人

The President of Tokyo Motor requests the pleasure of your company at a banquet held at the ABC Hotel.	東京車業的總裁邀請貴公司參加在 ABC 飯店舉辦的晚宴。
The pleasure of your company is cordially requested at Tokyo Motor Anniversary Celebration.	誠摯地邀請貴公司出席東京車業週年紀念慶祝會。
Cambridge Health Food Institute cordially invites you to attend our 30th Anniversary Party.	誠摯邀請您參加劍橋健康食品協會成立三十週年的紀念派對。

I write to invite you to attend our 2018 Annual Dinner on Friday, May 18, 2018 at the Ballroom of the Kodak Hotel.	我想邀請您參加於柯達飯店宴會廳所舉辦的二○一八年度晚會，時間為五月十八日。
The Board of Directors of XYZ Inc. cordially invites you to join its 45th Annual Customer Appreciation Banquet.	XYZ 企業董事會誠摯地邀請您參加第四十五屆年度酬賓宴會。
Stop by for some wine and cheese on Saturday, May 10 and help us celebrate the launch of our new products.	請於五月十日（星期六）前來品嚐酒飲與乳酪，並與我們一同慶祝新產品上市。
We would like to request the pleasure of your company at our Year End Party.	我們懇請貴公司光臨我們的年終宴會。
We request the pleasure of your company for cocktails.	我們邀請貴公司出席雞尾酒會。
Annual dinner is an occasion for us to thank our friends and supporters.	年終晚宴是為了感謝我們的友人與支持者而舉辦的。
You have always been a great support to us.	您總是大力支持我們。
You are invited to our 30th Anniversary Party.	歡迎您參加本公司的三十週年派對。
I would like to invite you to our formal dance party.	我希望能夠邀請您來參加本公司舉辦的正式舞會。
You are cordially invited to be our guest at the ceremony.	我們誠摯地邀請您參加此次典禮。
On May 20, we are giving a luncheon for Mr. Jonathan.	五月二十日，我們將為強納森先生舉辦一場午餐會。

宴會的資訊

The party will be held at the Royal Hotel in Tainan.	派對將於台南的皇家飯店舉辦。

Our reception starts at 6:30 p.m. and dinner will commence at 7:30 p.m.

宴會於六點半開始，七點半開始用晚餐。

Refreshments will be served from 2:30 p.m. to 5:00 p.m.

茶點供應時間為下午兩點半到五點。

Food and drinks will be available/ offered/on hand at that time.

到時將會提供餐點。

請求對方回覆

Would you please respond to this invitation by June 21?

請您於六月二十一日前回覆好嗎？

Would you please tell us if you could attend the party?

請告知您是否方便參加好嗎？

We would like to finalize our arrangement by June 20, please respond to this invitation by this Friday.

我們希望在六月二十日前安排內容，所以煩請於本週五前回覆。

Please e-mail me by the end of March, letting us know if you can attend.

請於三月底前告知您是否能出席。

I will call you on Thursday morning to confirm your decision.

我會於週四上午致電，與您確認。

信末再次邀請

We do hope that you will be able to join us in this occasion.

希望您能前來參與盛會。

Please come to help us celebrate this occasion.

請前來與我們一同慶祝。

Please come and celebrate with us!

請前來與我們同樂！

We request the honor of your company!

敬請光臨！

🖱 **I do hope you can join us for an evening of fun.**　　我希望您能夠與我們共度一個愉快的夜晚。

🖱 **Hope you have a good time in this party.**　　盼您在這場派對中玩得愉快。

 實力大補帖 **Let's learn more!**

 文法 / 句型解構

📍 邀請函常見用語

1 **R.S.V.P. = repondez s'il vous plait = please reply** 敬請回覆

ex. **Don't forget to R.S.V.P. before Wednesday.**
別忘了在週三前回覆。

2 **be cordially invited** 誠摯地邀請（邀請對象放「主詞」位置）

ex. **You are cordially invited to the wedding party on Sunday, June 18, 2018.**
誠摯地邀請您參加二〇一八年六月十八日（星期天）的婚禮派對。

3 **The event will be held in (place) at (time)** 將於（某時）於（某處）舉辦活動

ex. **The party will be held in the Royal Ballroom of Hilton Hotel at 7:00 p.m. on Friday, May 27, 2018.**
派對時間為二〇一八年五月二十七日（週五）晚上七點，將於希爾頓飯店的皇家宴會廳舉辦。

3-02 開幕邀請

From:	Lisa Miller
To:	all@mail.net
Subject:	Grand Opening Invitation

Dear Sirs:

Our new branch office in Taipei is opening on Friday, September 10, 2018.

We will have an office party on Friday night. You and your staff are cordially invited to attend our Grand Opening Party. It will be an informal affair with cocktails and snacks.

Please come and join us! We are looking forward to seeing you at the party!

Sincerely yours,

Lisa Miller

諸君：

我們的台北分公司即將於二〇一八年九月十日（星期五）開幕。

當晚會舉辦開幕派對，敬邀您與您的員工前來參加。此為非正式派對，我們會提供雞尾酒與點心。

請前來與我們同樂！希望能在派對上見到您！

麗莎・米勒 敬上

舉辦活動 & 邀請

Our new factory will be commencing production on June 20.

我們的新工廠將於六月二十日開始生產。

We are going to have an opening ceremony of our Taipei office.

我們將舉辦台北分公司的開幕典禮。

A luncheon will be held at the Hilton Hotel, followed by the opening ceremony and a tour of the new factory.

午宴將於希爾頓飯店舉行，隨後將是開幕典禮和參觀新工廠的行程。

There will be a banquet in the evening.

晚間將有一場宴會。

We request the pleasure of your presence at the opening ceremony of the company on Monday.

我們希望能有這個榮幸，邀請您前來參加星期一的開幕典禮。

We would like to invite you and your wife to be present at the celebration.

我們想邀請您與尊夫人一同前來參加慶祝活動。

We are inviting all firms that have contributed to our success.

我們邀請所有對我們的成功有貢獻的公司。

進一步的資訊

Attached is the map of our new office, which is located only 10 minutes on foot from MRT Shilin Station on the Tamsui line.

附件為辦公室地圖，新辦公室就位於淡水線士林站步行十分鐘的地方。

All arrangement for your stay overnight on that day will be made by us at our expense.

我方將支付您當天晚上的住宿費用。

信件結語：再邀請

We hope you can make an effort to attend our ceremony.	希望您能盡量出席。
The honor of your presence is requested.	敬邀您的光臨。
We kindly request your presence.	我們敬邀您的光臨。
I do hope that you will be able to spare the time to share this occasion with us.	希望您能撥冗參加我們的宴會。

實力大補帖 Let's learn more!

文法王 文法 / 句型解構

📍 邀請他人的句型

1 (Company Name) cordially invites you to... （公司名）誠摯地邀請您參加…。

2 I am writing to invite you to... 我寫信邀請您…。

3 We have pleasure in inviting you to... 很榮幸能邀請您…。

4 We would be honored if you could attend... 若您能出席…，我方將備感榮幸。

5 We would be pleased if you could join us at... 若您能參加…，我們將非常高興。

6 We would be delighted if you could come to… 若您能前來參加…，我們會很開心。

3-03 參加董事會議

邀請收件者參與會議的信件，一定要提醒對方「需要準備的事項」以及「會議重點」，以便讓對方提早準備。

From: Patrick Lee

To: Ms. Mason

Subject: The Board Meeting!

Dear Ms. Mason:

The Board of Directors of the Tokyo Motors Foundation will hold a meeting at 9 a.m., August 10, 2018 in our company.

On behalf of the Board, I am writing to invite you to join us and present your proposal for the new designed product. Then we would like to discuss the possible promotion strategies for the new product in the market.

We look forward to meeting you then.

Sincerely,

Patrick Lee

梅森小姐您好：

東京車業董事會將於二○一八年，八月十日早上九點於公司召開。

謹代表董事會，邀請您前來參加並提出新產品的方案，我們將於會中討論可行的產品促銷策略。

盼屆時與您見面。

派翠克・李 敬上

This is to inform you that our Board of Directors meeting will be held at meeting room A at 9 a.m., May 10, 2018.

特此告知，董事會將於二○一八年五月十日的上午九點，於 A 會議室召開。

I have the honor to call your attention to the agenda of the next meeting of the Board of Directors, scheduled on May 20, 2018.

很榮幸通知各位關於下一次董事會的議程，預計於二○一八年五月二十日召開。

The Board of Directors will be holding a meeting on May 20, 2018. We request all members to attend in person.

董事會將於二○一八年五月二十日舉行會議，所有人都必須出席。

Please find the agenda of the meeting in the attachment.

關於議程請見附件。

The most important point on the agenda is the discussion and adoption of the action plans of our merger next year.

本次議程最重要的目的，為討論明年合併案的行動方針。

The meeting will discuss the previously adopted motion and future action plans.

會議中將討論先前通過的議程，及後續的執行方針。

As the above matter requires due consideration and is of vital importance, your presence is required.

由於此次董事會議題相當重要，需要審慎思量，因此請您務必出席。

Please give your valuable suggestions at the meeting of the Board of Directors.

請於董事會上提供您寶貴的意見。

I hope you will attend the meeting without failure.

希望你能如期參加會議。

In anticipation of your great enthusiasm and kind cooperation!

期待您的熱情參與及合作！

3-04 邀請參加演講

在「邀請參加演講」的信件中，要積極地邀請對方，並簡單地介紹演講重點，以引起對方的注意。例如下文提到演講內容為今年癌症療法、會中可參與互動提問等（內容具體，對收件者而言才是有用的資訊）。

From: George Davis

To: peter.w@mail.net; ma55678@mail.net

Subject: Luncheon with Dr. Hoffmann

Dear Sirs:

We are giving a **luncheon**[1] for Dr. Hoffmann on June 6. Dr. Hoffmann will give a speech about the development of **medicine**[2] for **cancer**[3] **therapies**[4] in the **recent**[5] years.

During the speech, Dr. Hoffmann will be available to answer any questions you may have. We hope you will find this discussion interesting and **helpful**[6]. We hope you can attend.

The luncheon will be held at the Evergreen Hotel, Taichung at 12:00 p.m.

Kindly respond to this invitation by May 20.

Yours faithfully,

George Davis

諸君：

我們將於六月六日為霍夫曼博士舉辦午餐聚會。霍夫曼博士將於會中討論近幾年癌症療法的發展。

霍夫曼博士將於會中回答您所提出問題。希望您會覺得這個討論會有趣又實用，也期盼您當天的參與。

午餐聚會將於中午十二點整於台中長榮酒店舉辦。

請於五月二十日前回覆此邀請函。

喬治・戴維斯 敬上

E-mail 關鍵字一眼就通

① luncheon 名 午餐會　　② medicine 名 藥物　　③ cancer 名 癌症

④ therapy 名 療法　　⑤ recent 形 近來的　　⑥ helpful 形 有益的

複製 / 貼上萬用句 Copy & Paste

邀請參加演講

We believe you would be interested in the new iPhone 2018 seminar, which will be held on April 10.	我們相信您會對二〇一八年四月十日舉辦的新型 iPhone 研討會感興趣。
This Friday, we will be hosting 2018 International Conference-Economics, discussing "Challenges, Opportunities, and Strategies".	本週五，我們將舉辦二〇一八年的「國際經濟會議」，討論《經濟的挑戰、機會與策略》。
ABC Company invites you to a training of our latest software products.	ABC 公司邀請您來參加本公司最新軟體的培訓課程。
We are giving a speech at Holiday Hotel at 12 p.m. on this Friday to introduce our latest products.	我們將在本週五中午十二點，於假期飯店舉辦一場演講，介紹本公司的新產品。
You are invited to attend our annual company course this Friday.	我們邀請您本週五來參加我們公司一年一度的課程。
The presentation will take place at Holiday Hotel at 6 p.m. on March 8.	三月八日晚上六點，將於假期飯店舉辦一場發表會。
Are you available to attend the president's training this Saturday?	您是否能來參加本週六的總裁訓練課程呢？
Would you be willing to join the course to improve the skills and techniques of being the top salesman?	您是否願意來參加本課程，學會頂尖銷售員該具備的技巧與方法呢？

403

告知演講流程

🖱 **Dr. Midler will speak to us from 9:00 a.m. to 10:00 a.m.**

米德勒博士將於上午九點至十點發表演講。

🖱 **We will also have more paper presentations from 1:30 p.m. to 3:30 p.m.**

下午一點半至三點半，我們也將發表其他的論文。

🖱 **The training will be served from 2:00 p.m. to 5:00 p.m.**

這場訓練將在下午兩點至五點進行。

預先報名 & 繳費通知

🖱 **In order to reduce participants' costs, we negotiated to extend our deadline until this Friday for your participation.**

為了減少參加的費用，我們進行了協商，將截止日期延長到本週五，請踴躍參與。

🖱 **We hereby sincerely invite you again to register to this course.**

我們於此誠心地再次邀請您報名參加這個課程。

🖱 **Please simply reply to this e-mail with YES by Friday!**

請在週五前以「YES」回覆這封信件，告訴我們您要參加！

🖱 **Please let me know whether you can join us or not.**

請告知我您是否可以前來。

🖱 **Just call our office at 1881-9492 and we will be glad to secure a place for you.**

請致電我們辦公室：1881-9492，我們將樂意為您保留座位。

🖱 **Will you please come to the seminar?**

您是否可以出席研討會？

🖱 **I wonder when I can hear from you.**

不知您何時能決定？

🖱 **Would you please advise us if you could attend the seminar?**

請您通知我們是否可以參加好嗎？

🖱 **Please pay the tuition and fees by this week to secure your seat.**

為了保留您的席位，請於本週完成繳費。

The easiest way to wire is doing through ATM.	最簡單的方法，就是用自動提款機匯款。	**Part** **1** 入門篇
Please come to our office and pay the $8,000 course fee within this week (before Friday, Jan. 28) to secure your seat.	請於本週（一月二十八日，本週五前）至辦公室繳納八千元，以確保您的席位。	**Part** **2** 求職篇

信件結語

We will grant you a two-hour workshop certificate for this gathering.	我們將頒發這兩個小時的研討會證書給您。
We'll grant you an 8-hour conference certificate.	我們將頒發這八小時的會議證明給您。
Welcome to contact me with any questions.	若有任何問題，歡迎與我聯絡。
I will call you again on Friday morning to make sure everything is in order.	我會在週五上午再打電話給您，確認一切順利。
We are sure you will find the course interesting.	我們相信您會覺得這門課程很有趣。
Come join us!	請加入我們的行列！
I am looking forward to seeing you.	我期待與您相見。
We are looking forward to seeing you there.	我們期待在那兒見到您。
We hope you will be able to attend.	希望您能來參加。
I hope you'll make your time available and come to learn together with us.	希望您能撥冗和我們一起學習。
Please come and enjoy this conference together.	請和我們一起參與會議。

From:	Owen Smith
To:	Dr. Moretti
Subject:	Guest Speaker Invitation

Dear Dr. Moretti:

All our **staff**[1] of the Y&I Company have long **admired**[2] your excellent **research**[3] activities and would enjoy learning more about your work.

Therefore, we would like to ask you to be our guest speaker at a seminar to be held at the Hilton Hotel, Taipei, on November 23, 2018 at 10 a.m.

We would be very pleased if you could give a talk of about 50 minutes. After the talk, a 20-minute question and answer **period**[4] would be **ideal**[5]. About 50 persons are expected to attend the seminar. Most of them are the top representatives in our company.

All the traveling **expenses**[6] will be made by us. And we will pay an extra NT$10,000 to you. We would appreciate having your reply by October 30 so we can arrange the agenda.

We do hope you will be able to present on this occasion.

Yours faithfully,

Owen Smith

莫里蒂博士您好：

Y&I 公司的所有員工，長久以來一直欣賞您傑出的研究活動，並樂於向您學習。

因此，我們有意邀請您於十一月二十三日早上十點，於希爾頓飯店發表演講。

屆時希望您提供五十分鐘左右的演講。並期望能於演講後，安排二十分鐘的問答時間。參加研討會的人數約有五十人，其中大部分是我們的頂尖業務員。

我們會負擔所有的差旅費用，並將另外支付您一萬元新台幣。希望您能於十月三十日前

回覆本信，以便我們安排行程。

誠心企盼您能參加此次聚會。

歐文‧史密斯 敬上

Part
1
入門篇

Part
2
求職篇

Part
3
客戶往來篇

Part
4
人際互動篇

 E-mail 關鍵字一眼就通

1 staff 名 全體員工　　**2** admire 動 欽佩；欣賞　　**3** research 名 研究

4 period 名 期間　　**5** ideal 形 理想的　　**6** expense 名 花費

 複製 / 貼上萬用句 Copy & Paste

邀請來演講

Our company would like to invite you to speak at our annual seminar.	本公司希望能夠邀請您於年度研討會上發表演講。
The English Department of Kyoto University would like to extend to you an invitation to be our guest speaker at the annual conference.	京都大學英文系所特邀您出席學術年會，並作為特邀講者發表演說。
As you know, our department is interested in the 20th century English literature.	如您所知，本系所對於二十世紀的英國文學很有興趣。
Since you are very familiar with the field, we know that your view of points will be extremely interesting to us.	因為您是此領域的專家，所以您的見解肯定特別吸引人。
We hope you will be able to attend this conference and share your valuable experiences.	我們希望您能蒞臨此次會議，並分享您的寶貴經驗。

🖱️	I would be grateful if you could give a presentation on the topic of solar energy and solar panel.	若您能發表一篇介紹「太陽能與太陽能板」的演說，我將非常感激。
🖱️	The number of expected participants will be 1,000 people.	參加者預計有一千人。
🖱️	The participants will include nurses, doctors and the general public.	參加的人包含護士、醫師以及一般群眾。

演講相關資訊

🖱️	The seminar runs two hours, we would be glad if you could give an one-hour lecture.	研討會總共兩小時，我們希望您能提供一小時的講座。
🖱️	If you are not convenient at that scheduled time, we can reschedule for an earlier or later time.	如果這個時間您不方便，我們可以重新安排日程，將活動提前或延後。
🖱️	It would give us great pleasure if you could come to Taichung to give a talk to our English teachers.	如果您能蒞臨台中，為我們的英語教師發表演講，我們將備感榮幸。
🖱️	We can offer you a fee of NT$5,000 plus transportation and accommodation fees.	我們可以支付您新台幣五千元費用，以及交通費與住宿費。

信件結語

🖱️	Your participation will have great significance to our seminar.	您的出席將為我們這次的研討會帶來重要意義。
🖱️	We ask you to accept our sincere appreciation for your pleasant acceptance.	如您願意接受，我們會非常感謝。

🖱 Although we realize you are busy, we hope you can find time to accept the invitation.

我們知道您很忙碌，但仍希望您能撥冗參加。

🖱 You will receive further details later, but we would appreciate your acceptance soon so we may complete our agenda.

隨後您會收到相關細節，但希望您盡快給予答覆，以便我們安排行程。

✈ 實力大補帖 Let's learn more!

文法王 文法 / 句型解構

📍 演講可用句型

1 I want to talk about some of my work in the field of... 我想向各位報告關於我在…領域的工作成果。

2 I will give this talk in two parts. 我的演講將會分為兩部分。

3 The first part deals with XXX. And the second part concerns YYY. 第一部分討論的是 XXX；第二部分則進入 YYY 的內容。

4 I am afraid I won't have time to cover everything of.... 我恐怕沒有時間詳細說明…。

5 I will go through the next three points very briefly. 我將簡短說明下面三點。

深入學 資訊深度追蹤

📍 演講常見單字的比較

1 seminar：較常用於「教育相關」的會議，可能會包含訓練與知識交流。

2 conference：常見於商業領域和科學界，為「特定主題的正式會議」。

3 meeting：正式 / 非正式的會議。通常由一群人針對某主題討論，以得出解決方式。

4 workshop：工作坊，通常由一群具備共同興趣的人組成，互相訓練、學習。

3-06 參展邀請

邀請收信者參加活動時，除了詳細列出展覽訊息之外，若需要入場券才能參加，記得要提供索取入場券的方式、或是附於信中附上相關檔案。

From:	Judy Larsen
To:	Ms. Hansen
Subject:	Private Preview Showing - Summer Fashion Collection 2018

Private Preview Showing - Summer Fashion Collection 2018

Ticket No.55667

Dear Ms. Hansen:

As one of our longtime valued customers, we would like to invite you to our special Private Preview Showing of our Summer Fashion Collection for 2018.

The show will **take place**[1] at our **downtown**[2] store at 123 Beverly St., Los Angeles, Tuesday evening, July 6, 2018 from 2:00 p.m. to 6:00 p.m. Limited free parking will be available in our parking **garages**[3] on the Beverly Street side of the store.

For **entry**[4] into the show, you will be required to produce this original invitation with your ticket number printed on it. Therefore, please print this e-mail as your invitation ticket.

In order that we may plan for **snacks**[5] and **refreshments**[6] appropriately, if you plan to attend, please call Janet at 514-336-6517 and advise her by June 1.

Please note: If Janet doesn't hear from you by Friday, June 1, we will **assume**[7] that you are not attending the show and we will **issue**[8] your ticket number to others.

Part
1
入門篇

Part
2
求職篇

Part
3
客戶往來篇

Part
4
人際互動篇

Everyone here at The Fashion House looks forward to meeting you and sharing our Summer **Collection**[9] with you at our Private Preview Showing.

Sincerely yours,

Judy Larsen

非正式的預先展示秀 ─ 二〇一八年夏季時裝展

入場號碼：55667 號

韓森女士您好：

由於您是我們重要的客戶，因此，本公司敬邀您參加我們二〇一八年的私人夏季時裝展。

時裝秀地點在我們的城中店，位於洛杉磯比佛利街 123 號。時間則為二〇一八年七月六日，星期二下午兩點到六點。比佛利街上的停車場車位有限。

進入會場需要提供此封含有入場號碼的信函，因此，請您列印此郵件以便入場。

為確認需要準備的茶點數量， 請於六月一日前撥打 514-336-6517，來電告知珍妮特是否參加。

注意：如果珍妮特沒有在六月一日前收到您的回覆，即被視為不克前來，我們會將您的入場號提供給他人。

時尚屋的全體同仁期待您參與本次的夏季服裝預展。

茱蒂‧拉森 敬上

key words▲ E-mail 關鍵字一眼就通

1 **take place** 片 舉行　　2 **downtown** 形 商業區的　　3 **garage** 名 車庫

4 **entry** 名 進入　　5 **snack** 名 點心　　6 **refreshment** 名 茶點

7 **assume** 動 假定為　　8 **issue** 動 發給　　9 **collection** 名 收藏品

From:	Nick Wilson
To:	Mr. Thomas
Subject:	10th Annual Trade Show Invitation

Dear Mr. Thomas:

The 10th Annual Trade Show, featuring the most advanced 3C products, will be held in Taipei next month as follows. You are cordially invited to attend this most exciting event of this year.

Dates: March 1st to 5th, 2018.

Time: 9:00 a.m. to 6:00 p.m.

Place: Taipei World Trade Center Exhibition A

We will exhibit our latest products. Our expert staff will be on hand to provide you with all the information you need and answer any questions you may have. Free drinks and snacks will also be provided. You are welcome to bring along your friends.

For more details, please see our homepage http://www.e-world.com.tw/tradeshow. We look forward to seeing you at the show.

Best regards,

Nick Wilson

湯瑪斯先生您好：

第十屆年度商品展示秀將於下個月在台北展出，此次將展示最先進的 3C 產品，誠摯地邀請您參加今年最令人期待的盛會。

日期：二〇一八年三月一日至五日

時間：早上九點至下午六點

地點：台北世貿中心 A 展場

我們會展出最新商品，現場會有專人為您提供資訊，並為您解答疑惑。會場備有免費茶

點，歡迎攜伴參加。

若想進一步了解活動細節，請參考本公司網頁：http://www.e-world.com.tw/tradeshow，期待在展示會上看見您。

尼克‧威爾森 敬上

複製 / 貼上萬用句 Copy & Paste

信件開頭：參展資訊

The International Book Show will take place from July 7 to 10 in Taipei.	國際書展將在七月七日至十日於台北舉行。
The World Trade Fair falls on June 20 at the World Trade Center in Taipei.	世界貿易展覽會預計將於六月二十日在台北世界貿易中心舉行。
BCN Inc. invites you to an exclusive showing of its latest CD-ROM products.	BCN 公司邀請您來參加我們最新的光碟機產品展示會。
We would like to invite you to attend the exhibition, which our company will be participating.	我們想邀請您參加本次展示會，敝公司也將參展。
You are invited to a special showing of our new e-Books.	請來參加我們的新電子書展示會。
You are invited to a special showing of our new line of the new software.	誠摯地邀請您參加敝公司新軟體的特別發表會。
We are giving a luncheon at Grand Hyatt Hotel at 11 a.m. on Thursday to introduce our newest products.	我們即將於星期四上午十一點在君悅飯店舉辦午茶會，同時介紹最新產品。
The presentation will take place at the Evergreen Hotel at 3 p.m. on May 21.	說明會將於長榮酒店舉辦，時間為五月二十一日下午三點。

展覽會的相關事項

There are more than 500 exhibitors in attendance from around the world.	有超過五百家從世界各地前來的廠商參加展示會。
There are expected to be more than 100,000 visitors to attend the exhibition.	預估會有超過十萬人次前來參加展示會。
We are planning to exhibit our new software for designing.	我們將會展示最新的設計軟體。
We will demonstrate our new products.	我們會展示新產品。
Of course we will have our full line of other products on display, too.	我們當然也會展示其他商品。
Our staff will be there to assist you with any inquiries you may have during the exhibition.	展示會期間，我們的員工會在現場為您解答問題。
During the show, we are offering a special 10% discount on all our products.	在展出期間，我們提供全產品九折的優惠。
A special discount is only available once a year during the show period.	特殊優惠一年才一次，只有在展覽期間才有。

信件結語

You would benefit from a visit to see our best line of products.	您看了我們的產品後，會受益匪淺。
We hope that this exhibition will enable us to establish long-term business with the prospective clients.	希望此次參展，可與潛在顧客建立長期的商務關係。
For more information, please contact us by e-mail, or call 02-2233-4455.	如需更多資訊，請以電子郵件或電洽：02-2233-4455 聯絡。
For more details and an official invitation card, please see the attached files.	如需詳細資訊以及正式邀請函，請參看附件。

Part
1
入門篇

Part
2
求職篇

Part
3
客戶往來篇

Part
4
人際互動篇

3-07 接受邀請

若打算接受邀約時，只須簡單表明樂意參加之意，不需要太冗長的回信。

From:	Rose Murphy
To:	taylor0123@mail.net
Subject:	Re: 20th anniversary Invitation

Dear Sir:

President and Directors of Imperial Motors thank you for your kind invitation to the anniversary dinner party held at the Grand Hotel. They are pleased to attend.

Yours faithfully,

Rose Murphy

先生您好：

關於在圓山大飯店所舉辦的的週年慶祝晚宴，帝國車業的總裁及主管們皆感謝您誠心的邀約，他們樂於前往參加。

蘿絲‧墨菲 敬上

 複製 / 貼上萬用句 Copy & Paste

感謝對方的邀請

🖱 Thank you very much for your invitation.	謝謝您的邀請。
🖱 Thank you for remembering us.	謝謝您記得我們。
🖱 I would like to express my thanks.	在此表達本人的謝意。

It is really kind of you to include me in your event.	感謝您邀請我參與盛會。
Thank you very much for your kind invitation to the formal party on May 20.	非常感謝您邀請我們參加五月二十日的宴會。
I thank you for your kind invitation on the occasion of your opening ceremony.	謝謝您邀請我方參加貴公司的開幕典禮。
Many thanks for your kind invitation to the Grand Opening Ceremony.	非常感謝您邀請我們參加開幕典禮。
Thank you for inviting me to the dinner party.	謝謝您邀請我參加晚宴。
I accept the kind invitation for your anniversary dinner party with pleasure.	謝謝您誠心地邀請我參加週年晚宴。
We are pleased to receive your invitation and to participate in the seminar.	我們很開心接到研討會的邀請，也樂於參加。

答應參加

I am happy to attend your forum.	我很開心能參加論壇。
I am delighted to attend the party.	我很開心能參加聚會。
We are pleased to accept and will travel by flight arriving at Taipei at 5 p.m.	我們很開心受邀，當天會搭乘飛機於下午五點抵達台北。
We look forward to attending your opening ceremony.	我們很期待您的開幕典禮。
It would be an honor to attend your opening party.	能夠參加貴公司的開幕典禮是我的榮幸。
I am happy to attend the party.	我很樂意參加派對。
It is my pleasure to join you at the party.	能夠參與您的派對是我的榮幸。

🖱 Thank you for thinking of us during this special time. We're glad to join.	謝謝您在這個特別的時刻記得我們,我們很樂意參加。
🖱 I will be happy to be at your party at 7:00 p.m. on Saturday, June 21.	我很開心能夠參加您於六月二十一日,星期六晚上七點舉辦的派對。
🖱 I am very happy to attend your annual party.	我很開心參加貴公司的年度派對。
🖱 It is our honor to participate in the wonderful occasion.	能出席這樣美好的場合,我感到很榮幸。
🖱 I'm writing this letter to thank you for your kind invitation and to confirm my attendance.	在此感謝您的邀請,屆時我會出席。
🖱 I shall indeed be very happy to come, and look forward to meeting you with pleasure.	我很樂意參加,並且期待與您會面。
🖱 We look forward to sharing this joyful moment with you.	我們很期待分享您的歡樂時刻。
🖱 I will be there by 10 a.m.	我早上十點會抵達。

各種祝賀 & 肯定

🖱 Congratulations on your success in opening a new office!	恭喜您成立新的辦公室!
🖱 I understand how hard you have worked to reach this milestone.	我非常理解您付出了多少努力,才得以達到今日的成功。
🖱 We wish you every success for the future.	盼您未來事業蒸蒸日上。
🖱 We are taking this opportunity to wish your organization continued success.	利用這個機會,我們祝福貴公司業績長紅。

　　對於他人的邀約，若必須拒絕時，記得拿捏文字，使語氣委婉一點。建議於信中提供合理的理由，並於文末再次表達感謝之意與祝賀之語。

From:	Rose Murphy
To:	taylor0123@mail.net
Subject:	Re: 20th Anniversary Invitation

Dear Sir:

Mr. Liu, President of Imperial Motors, thanks you very much for the invitation to a dinner party at the Grand Hotel.

He would be delighted to accept; however, he has already made arrangements to attend another important meeting in Milan in June. Unfortunately, the meeting cannot be cancelled. Therefore, Mr. Liu will not be able to attend your dinner party.

Mr. Liu wishes to extend his cordial greetings to your President, Mr. Chang.

Yours faithfully,

Rose Murphy

先生您好：

帝國車業的總裁，劉先生感謝您邀請他參加在圓山大飯店的晚宴。

他很希望能夠參加；然而，他之前已安排好六月的行程，因此必須前往米蘭參加一場重要的會議。遺憾的是，這場會議無法取消。因此，劉先生無法參加晚宴。

劉先生希望能致上他誠心的祝福給貴公司的總裁，張先生。

蘿絲・墨菲 敬上

信件開頭：致歉

Part 1 入門篇

Part 2 求職篇

Part 3 客戶往來篇

Part 4 人際互動篇

I am really happy to receive your invitation to the annual dinner.	我很開心收到您年度餐會的邀請。
I would be obliged to accept, but unfortunately, I cannot attend the party owing to a prior appointment.	我很樂意參加，可惜的是，我先前已有其他邀約，因而無法參加。
I would like to attend, but I am afraid I will have to let you down.	我很想參加，但很抱歉要讓您失望了。
We would love to go, but we are afraid we couldn't make it.	雖然我們想參加，但我們恐怕無法前往。
I am sorry I am unable to join the forum in Taipei.	很抱歉，我無法參加這場在台北舉辦的研討會。
I am terribly sorry I will be unable to make appearance at your dinner party.	非常抱歉，我無法參加您的晚宴。
I have already made arrangements to attend another meeting, which cannot be cancelled.	我之前就已經安排要參加一場會議，這個行程無法取消。
We are sorry to inform you that we have to decline your invitation due to a prior engagement.	由於事先有約，所以無法應邀，非常抱歉。
Unfortunately, I cannot attend due to time conflict.	很不巧，由於時間不允許，恕我無法出席。
Regretfully, I am unable to join at that time.	很抱歉，當天我無法參加。
We regret that we are unable to accept the kind invitation of the party.	很抱歉，我們無法參加這場派對。
I greatly regret that I am unable to join you next Friday.	無法參與您下週五的聚會，我深感抱歉。

拒絕的原因 & 替代方案

We have to decline your invitation owing to a prior engagement.	由於事先有約，我不得不婉拒您的邀約。
Owing to a previous engagement, we are unable to accept the invitation.	由於我們已有其他約會，因此無法應邀前往。
I am scheduled to be in Europe in late June.	六月底我人在歐洲。
I would be out of town on business on that day.	我那天要出差，所以不會在城裡。
Unfortunately, I have to skip the party because of a prior appointment.	很不巧，因為我已和人有約，因而無法出席派對。
I regret that I am unable to accept because I have a prior engagement.	很抱歉我無法參加，因為事先已安排了其他的約會。
On behalf of me, Mr. Wu, manager of Sales Dept. of our company, will attend the party.	我們的業務部經理，吳先生會代表我參加派對。

信件結語

I hope you understand.	盼您諒解。
I do hope you will understand the reasons preventing my attendance.	我希望您能諒解我無法參加的理由。
Please accept my thanks again for inviting me.	我要再度感謝您的邀約。
Thank you again for including me.	再次感謝您邀請我。
I hope your party will be a great success.	祝您的派對圓滿成功。
We look forward to another such opportunity.	期待有下次機會。

Part
1
入門篇

Part
2
求職篇

Part
3
客戶往來篇

Part
4
人際互動篇

🖱 **Wish you a happy time. / Hope you have happy time.**　希望您玩得開心。

實力大補帖 Let's learn more!

文法王 ▲ 文法 / 句型解構

📍 拒絕開場白

1 I would be obliged to..., but unfortunately, ... 我很樂意…，可惜的是…。

2 I would like to..., but... 我很想…，但是…。

3 I am really sorry that... 我很抱歉，…。

4 I will be unable to... due to... 因為…，所以我無法…。

5 I cannot make it...because... 因為…，所以我無法…。

6 Regretfully to say, ... 很遺憾地，…。

深入學 ▲ 資訊深度追蹤

📍 寫拒絕信函的要點

　　拒絕信函要盡量委婉、禮貌，避免言語傷害到收件者。商業 E-mail 通常涉及與客戶的往來，所以必須特別注意，以免造成日後的麻煩或困擾。以下提示幾項重點，需要拒絕對方時不妨依此要點：

01 開頭先表示遺憾，語氣需真誠。

02 解釋拒絕對方的原因，盡量具體說明。

03 提出替代方案，讓被拒絕的一方有其他的選擇。

04 祝福對方未來一切順利。

4-01 感謝款待

　　為了出差或個人原因拜訪某地，若受到別人的招待，建議於事後寫信感謝提供招待的人，以表示禮貌。此類信件的內容無須太過冗長，但仍需正式。

From:	Willy Chang
To:	Ms. Davis
Subject:	Thank You for the Hospitality

Dear Ms. Davis:

Thank you very much for the warm help that I received during my visit in San Francisco. I greatly appreciate the generosity that was extended to me by everyone. Please convey my appreciation to all the members in your office.

Sincerely,

Willy Chang

戴維斯小姐您好：

前陣子拜訪舊金山，謝謝您的特別幫忙，也感謝大家對我的親切款待。請代我向貴公司的所有職員致謝。

張威利 敬上

信件開頭

On behalf of our entire delegation and myself, I thank you for your warm welcome.	謹代表本團體與我個人，感謝您的熱情款待。
Please convey my appreciation to your wife.	請替我向尊夫人致謝。
Thank you so much for the nice dinner and warm hospitality.	謝謝您提供的晚餐和熱情款待。
I wish to express my appreciation for the hospitality extended to me while I was in San Francisco last week.	上週在舊金山受到您的款待，在此致上感謝。
Please accept my warmest thanks for the hospitality extended to me during my stay.	致上最誠摯的感謝，謝謝您在我停留期間的熱情款待。
I would like to thank you for your great hospitality during my stay in London.	謝謝您在我停留倫敦期間的熱情款待。
I surely will not forget the hospitality I received during my stay there and kindness shown by you and your staff.	我絕對不會忘記在停留的期間，您與您的同仁是多麼熱情地招待我。
Thank you very much for showing me your factory and for the warm hospitality extended to me during my stay in Taichung last week.	上週在台中受到您的熱情款待，謝謝您帶我參觀工廠。
I would like to thank you again for the opportunity you gave me for visiting your office.	再次感謝您給我機會參觀您的辦公室。
I appreciate so much for your kindness in showing me around your new office.	謝謝您帶我參觀新的辦公室。

感謝對方的安排與幫助

I am pleased to have the chance to learn about the latest products from your factory.

我很開心有機會了解您工廠的新產品。

Thank you so much for a detailed presentation of your products.

謝謝您詳細地介紹產品。

Thank you for explaining to me the unique features of your company's products.

謝謝您向我解說貴公司產品的獨特之處。

Your presentation was not only very well organized, but also extremely interesting.

您的解說不僅條理分明，內容也十分有趣。

I learned a lot from my visit to your factory.

藉著本次參觀您工廠的機會，我學到很多。

My visit to America this time turned out to be very successful.

此次美國之行非常成功。

It was a stimulating experience for me.

對我來說，這是個令人興奮的經驗。

The new methods will certainly be useful in my work in Taiwan.

新的方法對我在台灣的工作很有助益。

I would like to thank you most sincerely for the efforts in organizing this meeting for me.

感謝您的大力相助，替我安排此次會議。

Thank you very much for your assistance.

謝謝您的幫忙。

信件結語

🖱	I look forward to meeting you and thank you personally when you come to Taiwan.	希望您來台灣時，我能當面謝謝您。
🖱	I really hope that you will have a chance to visit us in Taiwan.	希望您有機會拜訪台灣。
🖱	Hope to see you soon in Taiwan.	希望很快能在台灣與您見面。
🖱	Please do visit us when it is convenient for you.	要是您方便，一定要來讓我們招待。
🖱	I hope that someday you will be able to visit us.	希望有一天您也能來訪。
🖱	We should be glad to reciprocate for your kindness at any time.	希望能有機會回報您。
🖱	I hope I will have the opportunity reciprocatlng for your kindness in the near future.	希望不久之後就能有回報您的機會。

✈ 實力大補帖 Let's learn more!

活字典 單字 / 片語集中站

📍 表達感謝之情

be grateful for 感激	convey 傳達；表達
extend to 給予	hospitality 款待；好客
obliged 感激的	reciprocate 報答；回報

From:	Wendy Wood
To:	Amelia Huang
Subject:	Thanks for Your Congratulations

Dear Amelia:

Thank you very much for your warm greetings and the lovely flowers for my birthday, which were delivered this morning.

Please do come to visit us together with your husband when you are in England.

Best wishes to you and your family from my husband.

Best regards,

Wendy Wood

親愛的艾美利雅：

謝謝您對我的生日祝賀，還有今早送來的漂亮花朵。

如果你來英國，請務必和您先生一同來拜訪我們。

外子要我向您與家人問好。

溫蒂・伍德 敬上

複製 / 貼上萬用句 Copy & Paste

🖱 **I was delighted to receive your e-mail.**	收到您的電子郵件，我非常開心。
🖱 **Thank you for those good wishes.**	謝謝您的美好祝福。

🖱 Thank you so much for your congratulations.	非常感謝您的祝賀。
🖱 Thank you very much for your warm wishes for my birthday.	謝謝您送給我這麼溫暖的生日祝福。
🖱 Thank you so much for your congratulation e-mail.	謝謝您的電子祝賀郵件。
🖱 Your e-mail of congratulations gave us a great deal of pleasure.	您的電子祝賀郵件帶給我們莫大的喜悅。
🖱 Your e-mail of congratulations is deeply appreciated.	非常感謝您的電子祝賀郵件。
🖱 Many thanks for your congratulations and your lovely gift.	謝謝您的祝賀與可愛的禮物。
🖱 Thanks for choosing such a thoughtful gift.	謝謝你挑選了這麼貼心的禮物。
🖱 I am truly blessed to have you in my life.	我的生命中有你，真的很幸運。

✈ 實力大補帖 Let's learn more!

文法王 文法 / 句型解構

📍 表達感謝的句型

1 **I am writing this e-mail to express my sincerest appreciation for...**
藉此信表達我對…最誠摯的感謝。

2 **I can never thank you enough for...**
關於…這件事，我怎麼感謝你都不夠。

3 **I don't know how to thank you for...**
對於…，我真不知道要如何表達我內心的感謝之情。

4-03 感謝贈禮

收到贈禮後，記得要盡快寄出感謝函，才不會失禮。表達謝意時，除了明確表示你收到什麼禮物，讓對方確定禮物已交到你手上之外，還可以敘述這份禮物對你的意義，使送禮人與收禮人之間能互相傳遞情誼。

From: Allison Wood

To: Mr. Moore

Subject: Thank You for Your Present

Dear Mr. Moore:

I wanted to write you right away for the **lovely**[1] painting you sent to my office. Everyone here **appreciates**[2] your excellent **taste**[3] in **selecting**[4] the picture, which perfectly matches our office decor. You **have an eye for**[5] art.

We are so touched by your **thoughtfulness**[6]. I hope you will have the opportunity to visit our office in the near future. We look forward to seeing you soon in Sweden.

Best regards,

Allison Wood

摩爾先生您好：

感謝您送給我的畫，一收到禮物，我就迫不及待要寫信給您。辦公室裡的每個人看到畫，都覺得和公司的裝潢很搭，您對藝術真是獨具慧眼。

我們感謝您的貼心，希望您很快就有機會來訪，期盼近日能在瑞典見到您。

艾莉森・伍德 敬上

Part
1
入門篇

Part
2
求職篇

Part
3
客戶往來篇

Part
4
人際互動篇

key words ▲ E-mail 關鍵字一眼就通

1 lovely 形 美好的   **2** appreciate 動 欣賞   **3** taste 名 審美

4 select 動 挑選   **5** have an eye for 片 有眼光   **6** thoughtfulness 名 體貼

複製 / 貼上萬用句 Copy & Paste

致謝 & 表達喜愛

I want to take the time to write my thanks to you and ABC for the generous gift.	欲將感激寄於文字：感謝您和 ABC 公司送的大禮。
It was most thoughtful and generous of you to send this gift.	承蒙您致贈禮物，您真是體貼大方。
I would like to thank you for the beautiful present you sent us.	謝謝您寄給我們這麼漂亮的禮物。
Thank you for the lovely flowers you sent to me.	謝謝您寄來的美麗花朵。
I cannot find the words to thank you for sending me such a wonderful gift.	送給我這麼好的禮物，不知如何言謝。
Thank you very much for the gift; it was just what I wanted!	謝謝您的禮物；這正是我想要的！
It was something that I have always wanted.	這是我夢寐以求的東西。
I can never thank you enough for the thoughtful gift you sent to my son for his wedding.	真的很感謝您送給我兒子的結婚禮物。
The present looks beautiful in my living room and is a lovely reminder of the kind generosity of my friends and colleagues at XYZ.	禮物擺放在我的客廳裡看起來真棒，同時也提醒我在 XYZ 公司裡的朋友和同事們有多麼貼心。

🖱 Your gift brightened my day.	您的禮物讓我開心一整天。
🖱 Your thoughtfulness means so much to me.	您的貼心對我而言意義重大。
🖱 It is really a most acceptable gift which I will always treasure.	這真是最棒的禮物了，我會永遠珍藏。
🖱 I cannot find enough words to express my gratitude to you.	我找不出適當的言語來表達對您的感謝。
🖱 I don't know how to thank you for such an attractive present.	不知該如何感謝您送我這麼棒的禮物。
🖱 I feel moved by your affection.	您的好意讓我很感動。
🖱 Where did you find this present?	您是在哪找到這個禮物的？
🖱 It would be the most beautiful gift in the office.	這是辦公室裡最美的禮物。
🖱 I can't wait to show off the gift you gave me to my colleagues.	我等不及要拿禮物給同事看了。

信件結語：再次感謝

🖱 Thank you for thinking of me.	謝謝您想到我。
🖱 It was so nice to be remembered by you.	很開心您記得我。
🖱 Thank you, once again, for your gift and your kind thoughts.	再次感謝您的禮物和貼心。
🖱 Thank you again for the present.	再次感謝你送給我的禮物。
🖱 Thank you again for such a thoughtful gift.	再次感謝如此貼心的禮物。
🖱 I really appreciate your thoughtful present.	真的很感謝您送來這麼貼心的禮物。

Part
1
入門篇

Part
2
求職篇

Part
3
客戶往來篇

Part
4
人際互動篇

4-04 感謝協助

在「感謝協助」的信件中，可先說明感謝事由，再表達對收件者的感激之意。除此之外，還可以告訴對方，以後若有什麼幫得上忙的地方，你很樂意協助，以實際的回饋表達感謝，也是一種寫法。

From:	Lara Fischer
To:	Jessica Green
Subject:	Thanks for Help

Dear Jessica:

I am writing this to express thankfulness with all my heart.

Because of your suggestion and support, I have established my own company successfully. I do not know how to thank you enough. You really helped me a lot!

With warm regards,

Lara Fischer

親愛的潔西卡：

寫這封信是為了表達對您的感謝。

因為您的建議和支持，我成功創立了自己的公司。不管說什麼都不足以回饋您對我的幫助，您是我的恩人！

誠摯地祝福您。

拉娜・費雪 敬上

Allow me to thank you for your kind help.	感謝您的盛情相助。
Please accept our thanks for your timely help.	請接受我對您及時幫助的感激。
Please accept my sincere appreciation for accepting the trouble I had brought you.	請接受我誠摯的感謝，謝謝您這次的包容。
Allow me to give you my deep appreciation for your kind support.	請讓我感謝您的支持。
Please accept my sincerest thanks.	請接受我最誠摯的感謝。
I don't know how to express my appreciation for your valuable service.	您的幫助對我的助益非常大，真不知道該如何表達我的感謝。
Thank you for your tremendous help, which had helped us through the departmental restructuring.	謝謝您於部門重組時提供強大的協助，幫我們度過難關。
Thanks again for your assistance over the years.	謝謝您這幾年的協助。
I am writing this letter to thank you for your timely support during the time of crisis when I was helpless.	寫這封信是要感謝您在我無助時，及時拉了我一把。
Also, this isn't the first time you have helped me.	這已不是您第一次幫助我了。
You have always shown your compassion and care towards me.	您總是對我付出關懷。
Your timely support was crucial, and had made a difference in my hopeless life.	您當時的幫助至為關鍵，讓我本來毫無希望的生活有了完全不同的改變。

My life was in a hopeless situation where I received no help, and it was so until your timely support arrived.

直到得到您的幫助之前，我的生活一直很無助。

Your guidance and enormous trust in my skills have paved me the path to success.

您的指導和對我能力的信賴，讓我得以成功。

Without your assistance, I would not have finished it.

沒有您的協助，我不可能完成這件事。

I could never have made to this point without your greatest support.

若沒有你們的傾力相助，我絕不可能達成目標。

I would not have noticed it if you didn't remind me.

如果您沒有提醒我，我不會注意到的。

Without your reminder, I would not have solved the problem.

若沒有您的提醒，我就無法解決這個問題。

Thank you for the great advice.

謝謝您的建議。

I appreciate your support and guidance, which was really needed.

感謝您給予我所需要的支持以及指導。

I am grateful for all your help.

我很感激您的幫助。

Many thanks for your support.

謝謝您的支持。

I am so lucky having such a good friend like you.

我很幸運能擁有像您這樣的朋友。

Your support is really appreciated.

真的很感謝您的支持。

How can I repay you for such a great support?

我該怎麼回報您的支持呢？

I would like to thank you again for your help.

我要再次感謝您的幫助。

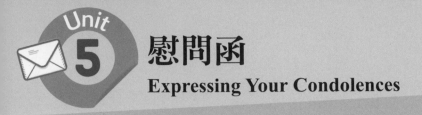
5-01 生病慰問

　　寫一封慰問信件給正在生病的人，可讓對方感受到溫暖與關懷。然而，倘若病情嚴重，甚至可能有不治之虞，就千萬不可以使用「早日康復」等不切實際的用詞。

From:	Helen Clark
To:	Joan Smith
Subject:	Get Well Soon

Dear Joan:

I hope you don't mind, but Tommy told me about your **illness**[1]. Please know that I'm always thinking about you and **pray**[2] for you. I'll give you a call soon, and hope I can **drop by**[3] for a visit after you get better.

Please also don't **hesitate**[4] to tell me if you don't want to be **bothered**[5]. I totally understand, and only hope for the best for you. I'm looking forward to seeing you as soon as you're **up to**[6] it.

Best regards,

Helen Clark

親愛的瓊安：

希望您不會介意湯米告訴我您生病的事，請知道我會掛念您，並為您禱告。我近日會致電給您，希望能在您覺得好一點時，前去拜訪。

如果您不希望被打擾，請告訴我，我不會見怪的。我完全能理解您的狀況，只希望您順利復原，那就最好了。期待見到您早日康復。

海倫・克拉克 敬上

1 illness 名 疾病 **2** pray 動 祈禱 **3** drop by 片 順道拜訪

4 hesitate 動 猶豫 **5** bother 動 打擾 **6** up to 片 勝任

複製 / 貼上萬用句 Copy & Paste

得知對方身體有恙

I have heard from a friend of mine that you are ill in bed.	我從友人口中得知您臥病在床。
I have learned with deepest regret that you are ill in bed.	獲知您臥病在床,我深感難過。
I was distressed to learn of your recent illness and hope that you are feeling better now.	得知您最近身體不適,希望您現在已經覺得好多了。
When your secretary told me that you would not be able to keep our appointment due to a sudden illness, I was deeply concerned.	當您的祕書通知我,您因為突然不適所以無法赴約時,我真的十分擔心。
I'm sorry to hear that you're not feeling well.	很遺憾聽到你生病的消息。
We are very sorry to hear that your recent illness has taken a bad turn.	獲知您的病情惡化,我們很難過。
I was shocked to hear that your mother had been hospitalized due to cancer.	我很驚訝令堂因為癌症住院。
I have heard from Janet that you are in the hospital.	我從珍妮特那裡得知您住院的消息。
I am sorry to hear that you have been hospitalized.	很遺憾聽到您住院的消息。

Due to health reasons, you have to go back to London.	因為健康因素，您必須回倫敦。
I am deeply concerned.	我深表關切。
David and I were devastated to learn about your son's illness.	聽到令郎生病的消息，大衛和我都十分震驚。
As parents, we do everything we can to protect our children.	身為父母，我們總是盡全力保護孩子。
I know how painful it would be for a parent to have children sick.	孩子生病，做為父母肯定會非常難受。

對方為住院病人

I am sure that the medical treatment you have received will be quite successful.	我相信您所接受的治療會非常成功。
We hope you feel comfortable and pleasant during your stay in the hospital.	希望您在住院期間，感到舒適且愉快。
I am sure you will have the best treatment that you need at the hospital, and hope this will lead to a speedy recovery.	我相信您在醫院裡會得到最好的治療，祝您很快就能康復。
If you are feeling up to meet with visitors, I will drop by to visit you.	如果您覺得好一點，可以見訪客的話，我會去看您的。
I will drop by and visit you as soon as you are allowed to have visitors.	您可以會客時，我就會去探望您。
I'm looking forward to seeing you as soon as you're up to it.	祝您早日復原，好讓我可以去探望您。

得知對方的復原情形良好

We are very happy to hear that you are making good progress in recovering.	很開心得知您的復原情形良好。

We are relieved to hear that you were not badly injured in the accident.	聽到您沒有受到太大的傷害，我們才放心。
I am glad your illness wasn't serious and that you will be soon out of the hospital and back with your family.	很高興您的病情並不嚴重，很快就可以出院回家。
We are happy to hear that he will be out in a few weeks.	我們很開心得知，他幾週後就能出院了。

信件結語

Please let me know if there is anything you need.	如果您需要任何幫忙，請讓我知道。
Everyone at the Taipei office and me wish you a speedy recovery.	台北分公司的同仁與我皆祝您早日康復。
Please accept my best wishes for a complete and speedy recovery.	希望您早日康復。
I trust that you are feeling better, and send you my best wishes for a speedy recovery.	我相信您已經好多了，祝您能迅速康復。
Get some more rest. I am sure you will have a full recovery soon.	多休息，你一定很快就會康復。
Take it easy and have a good rest!	放鬆心情，好好休養吧！
The business matter that we were going to discuss can be put off until you are back.	我們要討論的公事，可以等您回來之後再說。
Please know that you are in our thoughts and prayers.	請記得我們掛念著您，並為您祈禱。
You must stay strong, and don't forget that you are in my thoughts and prayers.	請保持堅強，並記得我永遠都會為你祈禱。

5-02 親友過世慰問

在哀悼信中，盡量不要刻意提及他人的痛楚或疾病，給予適度的安慰及鼓勵即可。

From: Ella Chou

To: Paul Smith

Subject: My Sincere Condolences

Dear Paul:

My **associates**[1] and I were deeply **sorrowful**[2] to learn that your father had **passed away**[3]. We all know that he will be greatly missed by everyone in your family. We wish to **convey**[4] our sincere **regret**[5] to members of your family.

Sincerely yours,

Ella Chou

親愛的保羅：

我與我的同僚對於令尊的過世感到非常哀慟，我們知道您的家人肯定都會非常想念他，我們謹向您的家人致上哀悼之情。

周艾拉 敬上

E-mail 關鍵字一眼就通

1 associate 名 同事   **2** sorrowful 形 悲傷的   **3** pass away 片 過世

4 convey 動 表達   **5** regret 名 哀悼

複製／貼上萬用句 Copy & Paste

聽聞消息

I was deeply sorrowful to hear of the sudden death of your mother.

聽到令堂驟逝的消息，令人深感哀慟。

I recently heard that you had a tremendous loss.

我剛聽聞您遭受重大損失。

I was shocked to hear of your mother's passing away.

聽到令堂辭世的消息，我非常震驚。

Please accept my condolences on the death of your beloved family member.

對於您家人的逝世，請接受我的哀悼。

I have just heard the loss of your father, and I hasten to send you my sincere condolences.

剛得知令尊過世的消息，在此致上我最誠摯的哀悼之意。

對逝者的讚嘆

Although I did not know him well, I know that he was a respected member in the community and the society.

雖然我不認識他，但我知道他在社會上備受尊崇。

His kindness, courtesy and experiences were rare qualities that we shall remember.

他寬厚仁慈、有禮、又閱歷豐富，這些高尚的品格，我們都會銘記在心。

Her passing must be a great loss to your family.

她的辭世一定是您家人莫大的損失。

Your mother, Lucy, was with such a kind and gentle soul.

令堂露西，有一顆善良溫柔的心。

She would do anything to help people in need.

她樂於幫助他人。

| Your father was always so considerate whenever I went to your house. | 每次我拜訪您時，令尊總是那麼體貼、周到。 |
| I shall always remember and appreciate the kindness of your mother. | 我會懷念令堂，並感謝她總是那麼仁慈地待我。 |

表達同理心

I can't imagine what a difficult time you're enduring.	我無法想像此刻您是多麼難以承受。
I'm truly sorry for your loss.	對於您所失去的，我感到非常遺憾。
I wish there is something I could say to soften your grief.	希望能說些安慰的話以減輕您的傷悲。
I know that words are not much comfort at time like this.	我知道在此刻，任何言語也不能安慰您。
Despite the fact that I truly know how you feel, no words of mine can ease your pain.	儘管我完全了解您的感受，但我明白沒有言語能夠安慰您。
We hope that you find courage and strength to bear this irreparable loss.	希望您有勇氣與能力承受這樣不可能挽回的損失。
May the love of those around you help you go through your difficult time!	希望您其他摯愛的親友可以幫您渡過難關。
I know how much you will miss your mother.	我知道您會非常思念令堂。
You are in my thoughts and prayers.	我會為您祈禱。
I want to let you know that even though I'm not physically there to help, I am here for you in your time of loss.	我想告訴您，雖然我無法親自幫忙，但我會在這段期間掛念著你。

Please let me know whether there is anything I can do to help during this difficult time.	在這樣艱難的時刻，如果有任何需要幫忙的地方，請讓我知道。
If there is anything we can do to help you during your grieving period, please know that we're here to help.	哀悼期間，若有任何需要，我們很樂意幫忙。
If you would like, I can come over tomorrow evening to babysit your kids.	如果您願意，我明天晚上可以替您照顧孩子。
I have some lovely pictures of Mary that I'd love to share.	我願意分享一些瑪莉的照片。

信件結語

I just want you know that you have my deepest and heartfelt concern.	我只是想讓您知道，我對此事深感遺憾。
We wish to extend our condolences to you and your family.	謹向您與您的家人致上哀悼之意。
I am writing to extend my deepest condolence to you and your family.	向您及家人致上我最深切的哀悼之意。
Our staff joins me in conveying our sincere condolences to members of your family.	本公司的同仁與我，希望向您的家人致上由衷的哀悼之意。
May God bless you and your family during this time and always.	願上帝保佑您與家人渡過此刻。
May the love of family and friends strengthen and support you for these days.	希望親友的愛與陪伴能使您堅強，度過這段時間。
I want to offer my deepest condolence.	我願致上我最深的哀悼。

6-01 公司開張

From:	Fiona Liu
To:	all@mail.net
Subject:	New Company Accouchement

Dear Sirs:

We would like to announce our new establishment of a trading firm under the name and address:

A&P Trading Co., Ltd.

111 Tien Yu Street, Taipei.

For more than 10 years, we have been the **representative**[1] for Tokyo Trading Co., Ltd. in Taiwan. We are confident that we have **considerable**[2] trading experiences in both Taiwan and Japan markets. We can offer you an exceptional large **variety of**[3] **first-class**[4] Taiwan computer **accessories**[5] at competitive prices.

Any information that you may require, we would be glad to **forward**[6] to you.

Yours faithfully,

Fiona Liu

Part
1
入門篇

Part
2
求職篇

Part
3
客戶往來篇

Part
4
人際互動篇

諸君：

在此通知：我們已成立一家貿易公司，公司的名稱與地址如下：

A&P 貿易股份有限公司

台北市天玉街 111 號

我們擔任日本貿易公司在台灣地區的代表超過十年，所以有非常多的台日貿易經驗。我們能提供品質一流、價格低廉的電腦周邊商品。

如果需要任何資訊，我們都很樂意提供。

費歐娜‧劉 敬上

key words ▲ E-mail 關鍵字一眼就通

1 representative 名 代表　　2 considerable 形 相當多的　　3 a variety of 片 種種

4 first-class 形 第一流的　　5 accessory 名 配件　　6 forward 動 遞送

複製 / 貼上萬用句 Copy & Paste

通知公司開幕

It is our pleasure to announce that we have started a new company in Tainan.	很開心通知您：我們在台南開了一家新公司。
I have the honor to inform you that we have just established a new company.	很榮幸通知您：我們開設了一家新公司。
I am very pleased to announce the opening of my footwear shop.	很高興宣布我的鞋店開幕了。
I am pleased to inform you that from this day, I shall commence the business of metal products.	很高興通知您：今後我將會經營有關金屬製品的公司。
We are pleased to inform you that due to the rapid increase of production, we have decided to open a new company.	由於產量遽增，我們決定開設另一家新公司。

其他資訊 & 結語

We have worked for many years to provide the best products and services in insurance and financing solutions.	我們有多年的經驗,將提供您保險財務方面的產品以及服務。
We would be delighted if you would take full advantage of our services and favorable shopping environment.	請善加利用本公司的服務與良好的購物環境。
The first 100 customers will become our VIPs.	前一百名客戶將成為我們的重要客戶。
Under the circumstances, we are sure that we will make you the fullest satisfaction.	在此情況下,我們確定您會十分滿意。

實力大補帖 Let's learn more!

深入學 資訊深度追蹤

♥ circumstance 與 environment 的比較

雖然這兩個單字都有表示「周圍環境」的意思,但意義卻差很多。

01 circumstance:表示「與某事件或某人有關的情況、環境等」。例句如:Under no circumstances will I believe you.(無論在什麼情形之下,我都不會相信你。)

02 environment:中文意思為「環境」,指「一個生物體或一群生物體周圍的整體狀況」。例句如:It is important to preserve the environment.(維護自然環境很重要。)

Part
1
入門篇

Part
2
求職篇

Part
3
客戶往來篇

Part
4
人際互動篇

6-02 公司合併

公司合併信件的內容，可提及合併的生效日、合併後是否會更改聯繫方式、以及合併對於公司的助益。

From:	Daniel Reid
To:	all@mail.net
Subject:	Notice of Merger

Dear Customers:

We would like to announce the merger of our company with NYE Co., **effective**[1] from May 1, 2018.

The new company will be called CYC Co. It will be **located**[2] at our present address. All telephone and fax number will **remain**[3] the same.

This merger represents a **pooling**[4] of the **expertise**[5] of two major cell phone firms, in an effort to develop and produce a **superior**[6] line of cell phones. It will also allow us to shorten our delivery time as we will have a much larger distribution network.

We hope the new company will provide you better service and quality products.

Best regards,

Daniel Reid

各位親愛的客戶：

在此通知：本公司即將與 NYE 公司合併，於二〇一八年五月一日生效。

新公司名稱為 CYC 公司，地址不變，所有的電話和傳真號碼也都不會變更。

兩家擁有專業手機技術公司的合併，主要用意為發展更優質的產品。除此之外，配送網路的擴大也將縮短我們的貨運時間。

希望新的公司能提供您更高品質的服務與產品。

丹尼爾・里德 敬上

1 effective 形 生效的　　**2** locate 動 設置在　　**3** remain 動 保持

4 pooling 名 企業聯營　　**5** expertise 名 專門技術　　**6** superior 形 較好的

🖋 複製 / 貼上萬用句 Copy & Paste

通知合併

As a businessperson, you know how important it is for a company to transform and grow in order to prosper.	身為一名生意人,您知道公司的轉型以及成功是多麼的重要。
NYE Company and ABC Company are pleased to announce a merger, which will take effect on May 1, 2018.	NYE 和 ABC 公司很開心宣布合併,於二〇一八年五月一日起生效。
AmBev Company is proud to announce that we have merged with Amoco Company to form AmBeo Company.	我們很驕傲地宣布:我們安貝夫公司即將與艾摩可公司合併,成為「安貝爾公司」。
From May 1, FOX Company will merge with our company to form Intercomm Company.	五月一日起,福斯公司將與我們公司合併,成立「音德康公司」。
The new company will be known as Bosch Company, with headquarters at 123 Tien-Yu St., Taipei.	合併後的公司名稱為「博世公司」,總公司設在台北市天玉街 123 號。
The new firm, DKSC Company's headquarters will be located at No.14 Tien-Yu St., Taipei.	新的 DKSC 總公司將位於台北市天玉街 14 號。

合併後的優勢

Our company title, contact information and commitment to quality will remain the same.

公司的名稱、聯絡方式和產品品質保證都會一樣。

This friendly merger will double the size of our staff and enable us to consolidate some of our back office functions.

這次的善意合併會使本公司的員工數加倍,並加強本公司的效能。

As a result of this merger, we are able to offer you a large range of business machines and electrical equipment.

公司合併後,將能提供您更大量的事務機器以及電子設備。

信件結語

In light of this move, customers should now look more favorably on the new company's larger scale.

公司的規模將會擴增,請各位以樂觀的態度看待此次合併。

We look forward to maintaining good relationship with our customers.

我們希望能繼續與客戶維持良好的商務關係。

Please give us an opportunity to supply you with our extended range of products.

請繼續給予我們為您提供產品的機會。

Stockholders with questions or who need assistance about the status of their shares may call 02-2272-3533.

股東如有疑問或需要協助,歡迎來電:02-2272-3533。

If you have any questions on this change in ABC Company, please do not hesitate to call our toll-free number.

如果您對 ABC 公司此次的合併有任何疑問,請撥打我們的免付費專線。

All of our customer service representatives will be happy to address any concerns you might have.

我們的客服人員將很樂意回答您的問題。

6-03 公司搬遷

在「公司搬遷通知」的信件中，記得要告知對方搬遷後的地點以及聯繫方式，以讓業務上的合作得以順利繼續。

From:	Lucy Scott
To:	all@mail.net
Subject:	Notice of Change of Address

Dear Customers:

This is to inform you that our office is moving to a new location as of May 31, 2018. Our new office address and telephone number are as follows:

Address: No.380 Chung Shan Road, Taichung

Telephone number: 04-5555-2233

Please find directions to our office on our website: www.sunnyside-trans.com.tw

Best regards,

Lucy Scott

親愛的客戶您好：

此函通知您：本公司將於二〇一八年五月三十一日遷往新的地點，新的地址與電話如下：

地址：台中市中山路 380 號

電話：04-5555-2233

詳細的位置資訊請見本公司網站：www.sunnyside-trans.com.tw

露西・史考特 敬上

複製 / 貼上萬用句 Copy & Paste

通知公司搬遷

We are glad to inform you that we have moved to a much larger and more convenient premises at:	此函通知您：本公司將搬移至空間更大、更方便的地點，新址如下：
We are pleased to announce that our office is moving to a new location.	本公司很高興宣布：我們將正式搬遷至新辦公室。
Please allow me to inform you that we have moved to the Chung-Shin Building in Shilin.	在此通知您：本公司已搬遷至士林的中興大樓。
Our warehouse will also be moved to our new location, making it easier to process your orders.	我們的工廠也將會搬遷至新址，以便處理您的訂單。
We have decided that a larger office is necessary to cope with the increasing demand of our products.	因應產品日漸增多的需求量，我們決定擴大辦公處。
We will relocate to a newer and larger place on February 1.	二月一日起，我們將搬遷至更大也更新的地點。
From September 10, our company will move to a new location.	九月十日起，本公司將搬至新地點。
The new address, phone number, and fax number are as follows:	以下為新的地址、電話以及傳真號碼：
Both the new telephone and fax number will be in operation from August 1.	新的電話和傳真將於八月一日開始運作。
Please address all correspondence to our new office as of July 31.	七月三十一日後，請將信件寄到本公司的新地址。
Please send all correspondence to us at:	請將信件寄到以下地址：

We are going to move into our new residence next Monday and will send you our new address.

我們預計下週一搬遷，稍後將告知您新的地址。

邀請對方來訪

The new premises include a customer's lounge.

新公司有一間客戶休息室。

Please feel free to call on at any time.

歡迎隨時來訪。

We are pleased if you will take the time to drop by our new office when you are around our area.

如果您到這附近，很歡迎您來參觀我們的新辦公室。

We would like to give you a tour of the facilities.

我們會帶您參觀新的辦公室。

We look forward to having you as a guest in our new office.

我們期待您來訪參觀新辦公室。

We will be pleased to have you visit our new office.

您若能來參觀新的辦公室，我們會很高興。

We would be happy to have you visit us if you are in our area.

如果您到附近，歡迎來訪指教。

其他 & 結語

We are celebrating our relocation to Taipei with a reception at the new office, from 7 to 10 p.m. on Friday, May 11.

我們將於五月十一日，星期五晚上七點到十點，於台北辦公室舉辦喬遷晚宴。

Please let me know if you could attend the party, I will send you an invitation.

如果您能來參加，請讓我知道，我將會寄發邀請函給您。

We look forward to continuing to provide you with quality service.

期盼今後仍能繼續提供您高品質的服務。

6-04 成立分公司

在「通知分公司成立」的信件中，除了提及成立時間，亦可簡述成立分公司之原因，如：因應市場需求、加快訂單處理速度及出貨速度等。信末可提供公司網址，讓對方得以點選連結，以了解更多資訊；或提供電子郵件信箱，供收件者直接回信。

From:	Benny White
To:	Mr. Thompson
Subject:	Opening of the New Branch in Paris

Dear Mr. Thompson:

We take pleasure in announcing the opening of the new branch in Paris! The new office will open on June 30 and our branch in London will be closed as of May 25.

You are cordially invited to our opening ceremony on June 30.

For more information, please send e-mail to admin@paris.com.

Best regards,

Benny White

湯普森先生您好：

我們很高興在此通知您：巴黎分公司即將開幕！新辦公室將於六月三十日啟用，因此，倫敦分公司將於五月二十五日結束營業。

我們誠心地邀請您參加六月三十日的開幕典禮。

欲了解更多資訊，請來信：admin@paris.com。

班尼・懷特 敬上

成立分公司

This is to inform you that we have opened a new branch in Edinburgh.	此函通知：我們已於愛丁堡成立分公司。
I am writing to inform you that our new office in Tokyo is ready to begin operations.	在此通知您：我們的東京分公司已準備好開始營運了。
We are glad to announce the opening of a new office in Taipei.	很開心通知您：我們的台北分公司即將開幕。
We are glad to announce that a new branch of our company in New York has been opened.	很開心通知您：我們的紐約分公司已經開幕了。
Next month, we shall be opening a Taichung branch.	我們下個月會成立台中分公司。
We wish to inform you that we have opened a new office in Paris.	我們想告知您：我們在巴黎成立新的辦公室了。
Due to the rapid demand from our customers, we are going to open another branch.	由於客戶的訂單量大幅增加，因此我們準備開設分公司。
Due to the popularity of our goods, we will be opening a new branch office in Tainan.	由於產品大受歡迎，因此我們即將在台南開設分公司。
As the demand of our trade in Taiwan is increasing, we will open a new branch in Kaohsiung next month.	由於在台灣的產品需求量增加，我們下個月會在高雄開設分公司。
Due to a constantly expanding demand for our products, we have decided to service our Taiwan customers with the new branch in Taichung.	由於持續成長的銷售量，我們已決定為台灣客戶成立台中分公司。

The new branch is located at the following address: | 新設分公司的地址如下：

優勢 & 開幕典禮

We would be able to process your orders more efficiently. | 我們處理訂單的效率將會提升。

We would deliver the goods much more quickly. | 我們寄送貨品的時間會加快。

We hope the new branch will be most helpful to our clients in the South Taiwan. | 希望新成立的分公司對我們南台灣的客戶有助益。

We would be pleased if you could attend our opening ceremony in Taipei. | 如果您能參加在台北舉辦的開幕典禮，我們會非常開心。

We are very honored by your attendance to our opening ceremony. | 我們非常榮幸您蒞臨我們的開幕典禮。

介紹負責人 & 信件結語

Our esteemed colleague, Ms. Eva Hung will be the new branch manager. | 我們尊敬的同仁，洪伊娃小姐將擔任新設分公司的經理。

The new branch manager has ten years of experiences in the market. | 新設分公司的經理擁有十年相關經驗。

Please feel free to contact our new branch manager. | 歡迎隨時與我們新上任的分公司經理連絡。

Thank you for your business and support of our products for a long time. | 謝謝您長久以來的合作與支持。

6-05 公司歇業

歇業通知函的內容須註明：一、日期；二、歇業理由；三、業務承接；四、感謝對方的長期配合。

From:	Mac Taylor
To:	all@mail.net
Subject:	Discontinuation of Business

Dear Customers:

We would like to inform you that our company will **discontinue**[1] business **operations**[2] as of January 20, 2019.

Because of the high interest rates, we are no longer able to provide the service that our customers require. We have made arrangements for all of our business **transactions**[3] to be handled by Fuji Company. They will be **contacting**[4] you in the near future to discuss the future **cooperation**[5] with you.

We would like to thank for past great business and would like to let you know how much we have enjoyed having you as our customers. We hope this does not cause you serious inconvenience.

Yours truly,

Mac Taylor

親愛的客戶：

在此通知您：本公司將於二〇一九年一月二十日結束營業。

由於利率偏高，本公司無法再提供客戶所要求的服務。我們已將後續業務交由富士公司處理，他們將於近日與您聯繫，並討論後續的商務關係。

很感謝您過去對本公司的支持，也很高興能為您服務。希望此事不會造成太大的不便。

麥克・泰勒 敬上

1 discontinue 動 停止　　**2** operation 名 營運　　**3** transaction 名 業務

4 contact 動 聯絡　　**5** cooperation 名 合作

✈ 實力大補帖 Let's learn more!

永久歇業

We would like to inform you that FOX Computers will be closing on July 31, 2018.	在此通知：福斯電腦公司將在二〇一八年七月三十一日結束營業。
We wish to inform you that we will discontinue all business operations from October 1.	此函通知：本公司即將於十月一日起結束營業。
This letter is a notification that due to financial difficulties, Sunshine Computers will regrettably be closed for business on August 20.	陽光電腦因為財務上的困難，將於八月二十日停止營業。
Due to circumstances beyond our control, we have decided not to continue with the business of Tour Company.	基於各種因素，我們決定中止旅遊公司的業務。
Our business will be discontinued after the end of August 1.	我們將於八月一日後結束營業。
We will be closing on July 10.	我們七月十日將結束營業。

暫停營運（日後會恢復營業）

This is to inform you that we are going to suspend business for some time.	謹以此信通知：本公司將暫停營業一段時間。

The restaurant will be closed during the holidays and reopen on Thursday, February 28.	本餐廳將於假期期間暫停營業，並於二月二十八日（星期四）重新開張。
Pleased be informed that our company will be closed temporarily from May 1st to 20th due to the building interior renovations.	在此通知：本公司因內部整修的關係，將於五月一日至二十日暫停營業。
We are informing you that XYZ Co. will be closed for the New Year's holidays from December 25 to January 4, and will reopen on Monday, January 5.	我們要通知您：XYZ 公司於十二月二十五日到一月四日的新年假期期間休業，並於一月五日重新開始營業。
We have made decision to upgrade the service equipment from today.	本公司決定從今日起，進行服務設施的升級作業。
Our company will be on holiday from May 1 to 5.	五月一日到五日，本公司將暫停營業。
We will reopen on March 21.	我們將於三月二十一日重新營業。

說明情況 & 後續處理

Our company is now on the verge of bankruptcy.	本公司目前面臨破產的危機。
Over the next two weeks, we will be reviewing our accounting records and paying any outstanding invoices.	接下來的兩週，我們會開始處理帳務並付清款項。
Over the next sixty days, we will be reviewing our accounting records and paying any outstanding invoices.	接下來的六十天內，我們會處理未付清的帳款。
If we have outstanding balance from your company, we will contact you to confirm our current obligation to your company so that we can settle accounts.	如果我們還有欠款，我們會聯絡您，並在了解款項後付清。

Our company will not place any additional orders to your company.	我們將不會再接受貴公司任何的訂單。
We will not be placing any additional orders with your company.	我們不會再向貴公司下訂單。
Arrangements have been made to ensure the smooth continuation of business operations.	有關後續的業務，我們已做好妥善的安排。
Next Monday, we are holding a clearance sale.	下週一，我們會舉辦清倉拍賣會。
Current stocks will be cleared regardless of cost.	庫存將不惜成本賣出。
Prices will be marked down by as much as on half.	所有商品皆半價出售。
As the sale is likely to be well attended, we hope you come as early as possible during the clearance sale.	此次特賣可能會吸引很多人，所以我們建議您儘早前來。

信件結語

Thank you very much for always supporting us.	謝謝您長年以來對本公司的支持。
Your continued support is highly appreciated.	感謝您長久以來的支持。
We are deeply sorry for any inconvenience it may cause.	對於造成的不便，我們深感抱歉。
Please don't hesitate to call if there is anything I can do for you before May 20.	在五月二十日之前，若您有任何問題，歡迎隨時來電。
We hope we will have the pleasure of doing business with you again in the future.	我們希望未來還有機會為各位服務。

From:	Laura Anderson
To:	Sally Walker
Subject:	Performance Improvement

Dear Sally:

We are pleased to have you in our department. Confirming the annual **feedback**[1] from your director, we **expect**[2] that you shall **improve**[3] your performance in the near future by:

1. Attending the meeting **on time**[4].

2. Finishing the report before the **due date**[5].

3. Cooperating with other colleagues.

We believe that you will be an excellent employee very soon by improving **above-mentioned**[6] points.

Yours sincerely,

Laura Anderson

親愛的莎莉：

我們對於你身為我們部門的一份子，感到很高興。從你主管口中得知你這一年的工作情況，我們希望你能在近期內改進以下幾點表現：

一、請準時參加會議。

二、在期限內完成報告。

三、請與同仁互相配合。

我們相信，只要改善上述事項，你很快就能成為一位優秀的員工。

蘿拉·安德森 敬上

1 feedback 名 回饋　　2 expect 動 期待　　3 improve 動 改善

4 on time 片 準時　　5 due date 片 期限　　6 above-mentioned 形 上述的

複製 / 貼上萬用句　Copy & Paste

We would like to tell you about certain unsatisfactory aspects of your performance in our company.	關於你的工作表現，有一些令人不甚滿意的情況，我們想要與你討論。
This is to inform you that your recent performance has failed to reach the targeted KPI.	謹以此信通知：你最近的工作績效沒有達到考核標準。
You are advised to take immediate actions to improve your performance as below points.	請針對以下幾點立即改善。
We expect that in the future you shall improve your performance by:	我們期待您在將來改善下述的工作表現：
Call the manager if you want to ask for sick leaves.	若要請病假，須打電話向經理報備。
Follow the regulations concluded by the meeting.	請遵守會議中討論出來的規定。
Clear your cubicle every week.	每週要整理自己的工作隔間。
Finish your work on time.	準時完成工作。
Arrive at the office on time.	準時上班。
Work harder.	更努力工作。
We have confidence the problem will not be repeated.	我們對你有信心，相信這些問題不會再發生。

459

筆記頁

筆記頁

筆記頁

國家圖書館出版品預行編目資料

10秒一貼不用抄！超人氣商用英文E-mail立可貼大全／
張翔 著. --初版. --新北市：知識工場出版 采舍國際有限
公司發行, 2017.09 面；公分· --（Mater；3）
ISBN 978-986-271-768-4（平裝）

1.商業書信　2.商業英文　3.商業應用文　4.電子郵件

493.6　　　　　　　　　　　　　　　106006542

知識工場・Mater 03

10秒一貼不用抄！超人氣
商用英文E-mail立可貼大全

出 版 者／全球華文聯合出版平台・知識工場
作　　者／張翔　　　　　　　印 行 者／知識工場
出版總監／王寶玲　　　　　　英文編輯／何牧蓉
總 編 輯／歐綾纖　　　　　　美術設計／吳佩真

台灣出版中心／新北市中和區中山路2段366巷10號10樓
電　　話／（02）2248-7896
傳　　真／（02）2248-7758
ISBN-13／978-986-271-768-4
出版日期／2021年最新版

全球華文市場總代理／采舍國際
地　　址／新北市中和區中山路2段366巷10號3樓
電　　話／（02）8245-8786
傳　　真／（02）8245-8718

港澳地區總經銷／和平圖書
地　　址／香港柴灣嘉業街12號百樂門大廈17樓
電　　話／（852）2804-6687
傳　　真／（852）2804-6409

全系列書系特約展示
新絲路網路書店
地　　址／新北市中和區中山路2段366巷10號10樓
電　　話／（02）8245-9896
傳　　真／（02）8245-8819
網　　址／www.silkbook.com

Knowledge is everything！

knowledge. 知識工場

Knowledge is everything！

nowledge. 知識工場

Knowledge is everything！

Knowledge is everything！